P9-ELR-104

# Our Inner Ape

Also by Frans de Waal

*Chimpanzee Politics* (1982)

*Peacemaking Among Primates* (1989)

*Good Natured* (1996)

*Bonobo* (1997)

*The Ape and the Sushi Master* (2001)

*My Family Album* (2003)

Frans de Waal

Photographs by the author

# Our Inner Ape

A LEADING PRIMATOLOGIST EXPLAINS

WHY WE ARE WHO WE ARE

RIVERHEAD BOOKS

a member of Penguin Group (USA) Inc.

New York

2005

RIVERHEAD BOOKS
Published by the Penguin Group
Penguin Group (USA) Inc., 375 Hudson Street, New York, New York 10014, USA • Penguin Group (Canada), 90 Eglinton Avenue East, Suite 700, Toronto, Ontario M4P 2Y3, Canada (a division of Pearson Penguin Canada Inc.) • Penguin Books Ltd, 80 Strand, London WC2R 0RL, England • Penguin Ireland, 25 St Stephen's Green, Dublin 2, Ireland (a division of Penguin Books Ltd) • Penguin Group (Australia), 250 Camberwell Road, Camberwell, Victoria 3124, Australia (a division of Pearson Australia Group Pty Ltd) • Penguin Books India Pvt Ltd, 11 Community Centre, Panchsheel Park, New Delhi–110 017, India • Penguin Group (NZ), Cnr Airborne and Rosedale Roads, Albany, Auckland 1310, New Zealand (a division of Pearson New Zealand Ltd) • Penguin Books (South Africa) (Pty) Ltd, 24 Sturdee Avenue, Rosebank, Johannesburg 2196, South Africa

Penguin Books Ltd, Registered Offices:
80 Strand, London WC2R 0RL, England


Published simultaneously in Canada

Library of Congress Cataloging-in-Publication Data

Waal, F. B. M. de (Frans B. M.), date.
Our inner ape : a leading primatologist explains why we are who we are / Frans de Waal; with photographs by the author.
p. cm.
Includes bibliographical references (pp. 239–257) and index.
ISBN 1-57322-312-3
1. Chimpanzees—Behavior. 2. Bonobo—Behavior. 3. Human behavior. 4. Psychology, Comparative. I. Title.
QL737.P96W3214 2005 2005042768
156—dc22

Printed in the United States of America
3 5 7 9 10 8 6 4 2

This book is printed on acid-free paper. ♾

*Book design by Michelle McMillian*

*For Cattie, my love*

# Contents

# Acknowledgments

This book owes so much to so many primates, human and nonhuman, that it's impossible to thank them all. The central idea was born from a discussion with Doug Abrams. At the time, I was thinking of applying my lifelong primate expertise to human behavior, and Doug felt that bonobos deserved far more attention than they had received thus far. The two ideas combined led to a book that directly compares human, chimpanzee, and bonobo behavior. Much more than my previous books, *Our Inner Ape* addresses our species' place in nature.

I appreciate feedback on my writing from Riverhead editor Jake Morrissey as well as Doug Abrams, Wendy Carlton, and my wife, Catherine Marin. I thank my agent, Michelle Tessler, for getting the book into such good hands.

Early in my career, in the Netherlands, I enjoyed the support of my adviser, Jan van Hooff, and the director of the Arnhem Zoo, Anton van Hooff, Jan's brother. I am grateful to Robert Goy for pulling me to this side of the Atlantic. In the United States, so many collaborators, technicians, and students worked with me that I won't mention names, but I owe all of them for their help with my studies and for opening fresh lines of inquiry. Finally, I thank Alexandre Arribas, Marietta Dindo, Michael Hammond, Milton Harris, Ernst Mayr, Toshisada Nishida, and Amy Parish for various forms of assistance, and Catherine for her love and support.

# Our Inner Ape

ONE

# Apes in the Family

One can take the ape out of the jungle, but not the jungle out of the ape.

This also applies to us, bipedal apes. Ever since our ancestors swung from tree to tree, life in small groups has been an obsession of ours. We can't get enough of politicians thumping their chests on television, soap opera stars who swing from tryst to tryst, and reality shows about who's in and who's out. It would be easy to make fun of all this primate behavior if not for the fact that our fellow simians take the pursuit of power and sex just as seriously as we do.

We share more with them than power and sex, though. Fellow-feeling and empathy are equally important, but they're rarely mentioned as part of our biological heritage. We would much rather blame nature for what we don't like in ourselves than credit it for what we do like. As Katharine Hepburn famously put it in *The African Queen*, "Nature, Mr. Allnut, is what we are put in this world to rise above."

This opinion is still very much with us. Of the millions of pages written over the centuries about human nature, none are as bleak as those of the last three decades—and none as wrong. We hear that we have selfish genes, that human goodness is a sham, and that we act morally only to impress others. But if all that people care about is their own good, why does a day-old baby cry when it hears another baby cry? This is how empathy starts. Not very sophisticated perhaps, but we can be sure that a newborn doesn't try to impress. We are born with impulses that draw us to others and that later in life make us care about them.

The old age of these impulses is evident from the behavior of our primate relatives. Truly remarkable is the bonobo, a little-known ape that is as close to us genetically as the chimpanzee. When a bonobo named Kuni saw a starling hit the glass of her enclosure at the Twycross Zoo in Great Britain, she went to comfort it. Picking up the stunned bird, Kuni gently set it on its feet. When it failed to move, she threw it a little, but the bird just fluttered. With the starling in hand, Kuni then climbed to the top of the tallest tree, wrapping her legs around the trunk so that she had both hands free to hold the bird. She carefully unfolded its wings and spread them wide, holding one wing between the fingers of each hand, before sending the bird like a little toy airplane out toward the barrier of her enclosure. But the bird fell short of freedom and landed on the bank of the moat. Kuni climbed down and stood watch over the starling for a long time, protecting it against a curious juvenile. By the end of the day, the recovered bird had flown off safely.

The way Kuni handled this bird was unlike anything she would have done to aid another ape. Instead of following some hardwired pattern of behavior, she tailored her assistance to the specific situation of an animal totally different from herself. The birds passing by her enclosure must have given her an idea of what help was needed. This kind of empathy is almost unheard of in animals since it rests on the ability to imagine the circumstances of another. Adam Smith, the pioneering economist, must have had actions like Kuni's in mind (though not performed by an ape) when, more than two centuries ago, he offered us the most enduring definition of empathy as "changing places in fancy with the sufferer."

The possibility that empathy is part of our primate heritage ought to make us happy, but we're not in the habit of embracing our nature. When people commit genocide, we call them "animals." But when they give to the poor, we praise them for being "humane." We like to claim the latter behavior for ourselves. It wasn't until an ape saved a member of our own species that there was a public awakening to the possibility of nonhuman humaneness. This happened on August 16, 1996, when an eight-year-old female gorilla named Binti Jua helped a three-year-old boy who had fallen eighteen feet into the primate exhibit at Chicago's Brookfield Zoo. Reacting immediately, Binti scooped up the boy and carried him to safety. She sat down on a log in a stream, cradling the boy in her lap, giving him a few gentle back pats before taking him to the waiting zoo staff. This simple act of sympathy, captured on video and shown around the world, touched many hearts, and Binti was hailed as a heroine. It was the first time in U.S. history that an ape figured in the speeches of leading politicians, who held her up as a model of compassion.

## THE HUMAN JANUS HEAD

That Binti's behavior caused such surprise among humans says a lot about the way animals are depicted in the media. She really did nothing unusual, or at least nothing an ape wouldn't do for any juvenile of her own species. While recent nature documentaries focus on ferocious beasts (or the macho men who wrestle them to the ground), I think it's vital to convey the true breadth and depth of our connection with nature. This book explores the fascinating and frightening parallels between primate behavior and our own, with equal regard for the good, the bad, and the ugly.

We are blessed with two close primate relatives to study, and they are as different as night and day. One is a gruff-looking, ambitious character with anger-management issues. The other is an egalitarian proponent of a free-spirited lifestyle. Everyone has heard of the chimpanzee, known to science since the seventeenth century. Its hierarchical and murderous behavior has inspired the common view of humans as "killer apes." It's our biological destiny, some scientists say, to grab power by vanquishing others and to wage war into perpetuity. I have witnessed enough bloodshed among chimpanzees to agree that they have a violent streak. But we shouldn't ignore our other close relative, the bonobo, discovered only last century. Bonobos are a happy-go-lucky bunch with healthy sexual appetites. Peaceful by nature, they belie the notion that ours is a purely bloodthirsty lineage.

It is empathy that allows bonobos to understand each other's needs and desires and to help achieve them. When the two-year-old daughter of a bonobo named Linda whimpered at her mother with pouted lips, it meant that she wanted to nurse. But this infant had been in the San Diego Zoo's nursery and was returned to the group long after Linda's milk had dried up. The mother understood, though, and went to the fountain to suck her mouth full of water. She then sat in front of her daughter and puckered her lips so that the infant could drink from them. Linda repeated her trip to the fountain three times until her daughter was satisfied.

We adore such behavior—which is itself a case of empathy. But the same capacity to understand others also makes it possible to hurt them deliberately. Both sympathy and cruelty rely on the ability to imagine how one's own behavior affects others. Small-brained animals, such as sharks, certainly can hurt others, but they do so without the slightest idea of what others may feel. The brains of apes, on the other hand, are one-third the size of ours, making them sufficiently complex for cruelty. Like boys throwing rocks at ducks in a pond, apes sometimes inflict pain for fun. In one game, juvenile lab chimpanzees enticed chickens behind a fence with bread crumbs. Each time the gullible chickens approached, the chimps hit them with a stick or poked them with a sharp piece of wire. This Tantalus game, which the chickens were stupid enough to play along with (although we can be sure it was no game to them), was invented by the chimps to fight boredom. They refined it to the point that one ape would be the baiter, another the hit man.

Apes are so like us that they're known as "anthropoids," from the Latin for "humanlike." To have two close relations with strikingly different societies is extraordinarily instructive. The power-hungry and brutal chimp contrasts with the peace-loving and erotic bonobo—a kind of Dr. Jekyll and Mr. Hyde. Our own nature is an uneasy marriage of the two. Our dark side is painfully obvious: An estimated 160 million people in the twentieth century alone lost their lives to war, genocide, and political oppression—all due to the human capacity for brutality. Even more chilling than such incomprehensible numbers are more personal expressions of human cruelty, such as the appalling incident in a small Texas town in 1998 in which three white men offered a forty-nine-year-old black man a ride. Instead of taking him home, they drove him to a deserted spot and beat him, then tied him to their truck and dragged him several miles along an asphalt road, tearing off his head and right arm.

We are capable of such savagery despite, or perhaps precisely *because* of, our ability to imagine what others feel. On the other hand, when that same ability is combined with a positive attitude, it prompts us to send food to starving people, make valiant efforts to rescue complete strangers (such as during earthquakes and fires), cry when someone tells a sad story, or join a search party when a neighbor's child is missing. With both cruel and compassionate sides, we stand in the world like a Janus head, our two faces looking in opposite directions. This can confuse us to the point that we sometimes oversimplify who we are. We either claim to be the "crown of creation" or depict ourselves as the only true villains.

Why not accept that we are both? These two aspects of our species correspond to those of our closest living relatives. The chimpanzee demonstrates the violent side of human nature so well that few scientists write about any other side at all. But we are also intensely social creatures who rely on one another and actually need interaction with other people to lead sane and happy lives. Next to death, solitary confinement is our most extreme punishment. Our bodies and minds are not designed for lonely lives. We become hopelessly depressed in the absence of human company, and our health deteriorates. In one recent medical study, healthy volunteers exposed to cold and flu viruses got sick more easily if they had fewer friends and family around them.

This need for connection is naturally understood by women. In mammals, parental care cannot be separated from lactation. During the 180 million years of mammalian evolution, females who responded to their offspring's needs outreproduced those who were cold and distant. Having descended from a long line of mothers who nursed, fed, cleaned, carried, comforted, and defended their young, we should not be surprised by gender differences in human empathy. They appear well before socialization: The first sign of empathy—crying when another baby cries—is already more typical in girl babies than boy babies. And later in life empathy remains more developed in females than in males. This is not to say that men lack empathy or don't need to connect with others, but they seek it more from women than from other men. A long-term relationship with a woman, such as marriage, is the most effective way for a man to add years to his life. The flip side of this picture is autism—an empathy disorder that keeps us from connecting with others—which is four times more common in males than females.

The empathic bonobos regularly put themselves into someone else's shoes. At the Georgia State University Language Research Center in Atlanta, a bonobo called Kanzi has been trained to communicate with people. He has become a bonobo celebrity, known for his fabulous understanding of spoken English. Realizing that some of his fellow apes do not have the same training, Kanzi occasionally adopts the role of teacher. He once sat next to Tamuli, a younger sister who has had minimal exposure to human speech, while a researcher tried to get Tamuli to respond to simple verbal requests; the untrained bonobo didn't respond. As the researcher addressed Tamuli, it was Kanzi who began to act out the meanings. When Tamuli was asked to groom Kanzi, he took her hand and placed it under his chin, squeezing it between his chin and chest. In this position, Kanzi stared into Tamuli's eyes with what people interpreted as a questioning gaze. When Kanzi repeated the action, the young female rested her fingers on his chest as if wondering what to do.

Kanzi understands perfectly well whether commands are intended for him or for others. He was not carrying out a command intended for Tamuli—he actually tried to make her understand. Kanzi's sensitivity to his sister's lack of knowledge, and his kindness in teaching her, suggest a level of empathy found, as far as we know, only in humans and apes.

## WHAT'S IN A NAME?

In 1978, I first saw bonobos up close at a Dutch zoo. The label on the cage identified them as "pygmy chimpanzees," implying they were just a smaller version of their better-known cousins. But nothing could be further from the truth.

A bonobo is physically as different from a chimpanzee as a Concorde is from a Boeing 747. Even chimps would have to admit that the bonobo has more style. A bonobo's body is graceful and elegant, with piano-player hands and a relatively small head. The bonobo has a flatter, more open face with a higher forehead than the chimpanzee. A bonobo's face is black, its lips are pink, its ears small, and its nostrils wide. Females have breasts; they are not as prominent as in our species, but definitely A-cup compared to the flat-chested other apes. Topping it all off is the bonobo's trademark hairstyle: long black hair neatly parted in the middle.

The biggest difference between the two apes is body proportion. Chimps have large heads, thick necks, and broad shoulders—they look as if they work out in the gym every day. Bonobos have a more intellectual appearance, with slim upper bodies, narrow shoulders, and thin necks. A lot of their weight is in their legs, which are longer than a chimp's. The result is that when knuckle-walking on all fours, the chimp's back slopes down from powerful shoulders, whereas the bonobo's remains fairly horizontal because of its elevated hips. When standing or walking upright, a bonobo seems to straighten its back better than a chimp, giving the bonobo an eerily humanlike posture. For this reason, bonobos have been compared to "Lucy," our *Australopithecus* ancestor.

The bonobo is one of the last large mammals to be discovered by science. The discovery took place in 1929, not in a lush African habitat, but in a colonial Belgian museum following the inspection of a small skull thought to have belonged to a juvenile chimp. In immature animals, however, the sutures between skull bones ought to be separated. In this skull, they were fused. Concluding that it must have belonged to an adult chimp with an unusually small head, Ernst Schwarz, a German anatomist, declared that he had stumbled upon a new subspecies. Soon the anatomical differences were considered important enough to elevate the bonobo to the status of an entirely new species: *Pan paniscus.*

A biologist who had been a student with Schwarz in Berlin told me how Schwarz's peers used to make fun of him. Schwarz not only claimed there were two chimp species, but also that there were three elephant species. Everyone knew that there was only one of the first and two of the second. Their standard line about *der Schwarz* was that he knew "everything and more." As it turns out, Schwarz was right. The African forest elephant was recently confirmed as a separate species, and Schwarz is known as the official discoverer of the bonobo—the sort of honor scientists are willing to die for.

The bonobo's genus name, *Pan,* derives appropriately enough from the Greek forest god with a human torso and the legs, ears, and horns of a goat. Playfully lecherous, Pan loves to frolic with nymphs while playing the shepherd's (or pan) flute. The chimpanzee belongs to the same genus. The bonobo's species name, *paniscus*, means "diminutive," whereas the chimp's species name, *troglodytes*, means "cave dweller." With the bonobo being called a small goat deity and the chimp a grotto goat deity, these are curious epithets indeed.

The name "bonobo" probably derives from a misspelling on a shipping crate from "Bolobo," a town on the Congo River (although I have also heard that "bonobo" means "ancestor" in an extinct Bantu language). In any case, the name has a happy ring to it that befits the animal's nature. Primatologists jokingly employ it as a verb, as in "We're gonna bonobo tonight," the meaning of which will soon become clear. The French refer to bonobos as "Left Bank chimpanzees"—a name that summons up images of an alternative lifestyle—since they live on the south bank of the westward streaming Congo River. This mighty river, which in places is ten miles wide, permanently separates bonobos from the chimpanzee and gorilla populations to the north. Despite the bonobo's previous name, "pygmy chimpanzee," they are not much smaller than chimpanzees. The average adult male bonobo weighs ninety-five pounds and the average female eighty pounds.

What struck me most while watching my first bonobos was how sensitive they seemed. I also discovered some habits that shocked me. I witnessed a minor squabble over a cardboard box, in which a male and female ran around and pummeled each other until all of a sudden the fight was over and they were making love! I had been studying chimps, which never switch so easily from fighting to sex. I thought this bonobo behavior was an anomaly or that I had missed something that could explain the sudden change of heart. But it turned out that what I had seen was perfectly normal for these Kama Sutra primates.

I learned this much later, after I had begun working with bonobos at the San Diego Zoo. Information on wild bonobos had been trickling in over the years from Africa, which added to our knowledge about this mysterious cousin. Bonobos are native to a relatively small area, about the size of

England, in the Democratic Republic of the Congo (formerly Zaire) where they live in dense, humid swamp forest. When they enter a clearing where field workers leave sugarcane, the males arrive first. They hurry to collect handfuls of stalks so that they have their share before any females show up. When the females arrive, their entrance is accompanied by lots of sex among everybody and the inevitable appropriation of the best food by the older matriarchs. The same is true for the zoo colonies that I've studied, which are invariably ruled by an older female. This is surprising given that male and female bonobos differ as much in size as humans do, with the average female bonobo weighing 85 percent as much as the average male. On top of this, male bonobos have sharp canine teeth, which the females lack.

So how do the females maintain control? The answer is solidarity. Take Vernon, a male bonobo at the San Diego Zoo, who used to rule a small group that included one female, Loretta, who was his mate and friend. This was the only instance I have ever seen of a bonobo group run by a male. At the time, I thought this was normal: after all, male dominance is typical of most mammals. But Loretta was relatively young, and she was the only female. As soon as a second female was added to the group, the power balance shifted.

The first thing Loretta and the other female did upon meeting was have sex. The pattern is known by experts as genito-genital- (or GG-) rubbing, but I have also heard it called, more colorfully, the "hoka-hoka." One female wrapped her arms and legs around the other and clung to her the way an infant bonobo attaches itself to its mother's belly. Facing each other, the two then pressed their vulvas and clitorises together, rubbing them sideways in a rapid rhythm. They had big grins on their faces and squealed loudly, leaving little doubt about whether apes know sexual pleasure.

Sex between Loretta and her new female friend became more and more common, spelling the end of Vernon's rule. Months later the typical scene at feeding time was the females having sex and then together claiming all the food. The only way for Vernon to get any food was by begging with outstretched hand. This is also typical among wild bonobos, where females control the food supply.

Compared with the male-centered chimpanzee, the female-centered, erotic, and peaceable bonobo offers a fresh way of thinking about human ancestry. Its behavior is hardly consistent with the popular image of our forefathers as bearded cavemen dragging women around by their hair. Not that things were necessarily the other way around, but it is good to be clear on what we know and don't know. Behavior doesn't fossilize. This is why speculations about human prehistory are often based on what we know about other primates. Their behavior indicates the range of behavior our ancestors may have shown. And the more we learn about bonobos, the more this range expands.

## MAMA'S BOYS

Not too long ago, I spent a typical day at the San Diego Zoo with two old friends, Gale Foland and Mike Hammond, both veteran ape keepers. This is not a job for everyone. It's impossible to deal with the needs and reactions of apes without tapping the same emotional reservoir we use to deal with fellow human beings. Keepers who fail to take apes seriously will never get along with them, and those who take them too seriously will succumb to the web of intrigues, provocations, and emotional blackmail that suffuses every ape group.

In an area away from the public, we leaned over a balustrade looking down on a spacious, grassy enclosure. The air carried the unique pungent smell of gorillas. Earlier this morning, Gale had introduced a five-year-old female named Azizi, whom he had raised himself, to the enclosure. Azizi found herself in a group with a new male, Paul Donn, a gargantuan character who is leaning against the wall. Occasionally, he charges around the enclosure, beating his chest, to impress the gathering of females that he is in control, or at least would like to be. Female gorillas, especially the older ones, tend to disagree: they sometimes band together to chase him around "to keep him in line," as Gale says. But for the moment Paul Donn is calm, and we see Azizi shuffle closer and closer toward him. The male acts as if he fails to notice, tactfully inspecting his toes and never looking straight at the nervous gorilla girl. Each time Azizi gets a little bit nearer, she glances up at Gale, her adoptive parent. She locks her eyes into his. Gale nods and says such things as "Go on, don't be scared." This is easy for him to say: Paul Donn is probably five times Azizi's weight, all of it muscles. But Azizi is irresistibly attracted.

These gorillas are known for their intelligence. Gorillas are not supposed to use tools: in the wild they never do. But three gorillas at the zoo have found a new way to get at the tasty leaves of the fig trees. They are prevented from climbing the trees by hot wire, but get around this by picking up one of the many branches lying around, standing up on two legs and flinging it into a tree. The branch usually comes back down with some of the foliage. One female was seen breaking a long stick in two and using the more suitable piece—an important step, as it showed that gorillas are capable of modifying their tools.

Today, an incident takes place involving the same hot wire. It's the sort of scene that catches my eye. One older resident female has learned to reach under the wire without getting shocked to feed on weeds that grow beyond the wire. Right next to her sits a new female, who, Gale tells me, has just recently been shocked for the first time. It was a nasty experience for her, resulting in screaming and frantic handshaking. The new female has befriended the other and now sits watching her do exactly what had caused her so much pain. As soon as she sees her friend reach under the wire, she jumps behind her and starts pulling at her. She wraps an arm around her middle and tries to move her away from the electric fence. But her older friend doesn't budge; instead, she continues to reach. After some time, the new female sits back, watching intently, while wrapping both of her arms tightly around herself. She seems to be bracing herself for the shock she believes the other will get. "Changing places in fancy" indeed.

Like chimpanzees and bonobos, gorillas are known as great apes. There are only four great ape species, the fourth one being the orangutan. Apes are primates that are large and lacking tails. Both traits set the family of humans and apes—known as the Hominoids—apart from the monkeys. Apes should therefore never be confused with monkeys (there is no better way to insult an ape expert than to say that you love his monkeys!), and "primate" is the more inclusive label, as it also applies to us. Among the apes, our closest kin are chimps and bonobos, neither one of which is closer to us than the other. Yet this doesn't prevent primatologists from hotly debating which one is the best ancestral human model. We all derive from a single forebear, and one species may have retained more of its traits than the other, making it more relevant to human evolution. But it's impossible right now to determine which one. Not surprisingly, chimp experts generally vote for the chimpanzee and bonobo experts for the bonobo.

Since gorillas split off from our evolutionary branch slightly before chimps and bonobos, it's been argued that whichever ape is more like the gorilla deserves to be called the original type. But who says that gorillas themselves resemble our last common ancestor? They have had plenty of time to change, too—over seven million years, in fact. What we are looking for instead is the ape that has changed the least over time. Takayoshi Kano, the premier expert on wild bonobos, has argued that since bonobos never left the humid jungle—whereas chimpanzees did so partially and our own ancestors completely—bonobos probably have encountered fewer pressures to change and may therefore look most like the forest ape from which we all descend. American anatomist Harold Coolidge famously speculated

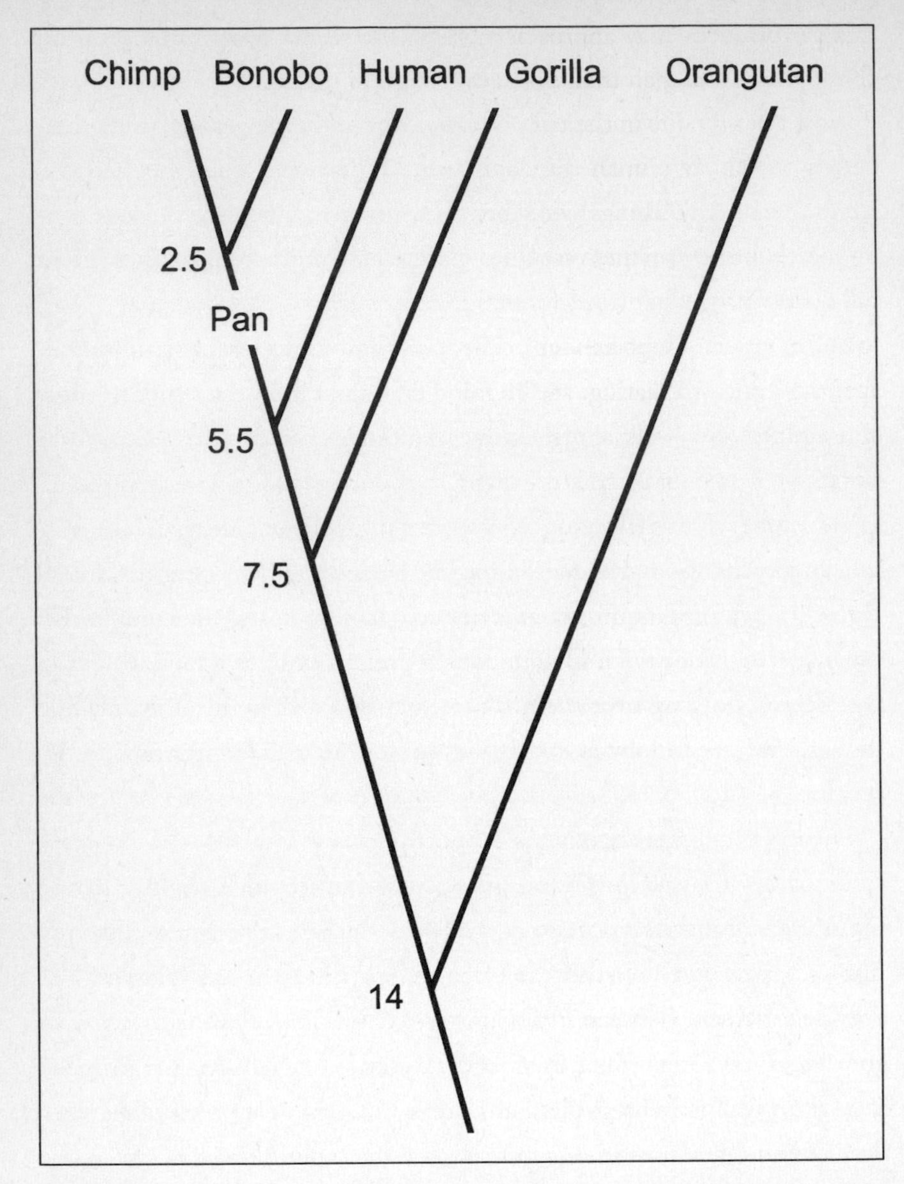


Tree of origin of humans and the four great apes based on DNA comparisons. Data points indicate how many millions of years ago species diverged. Chimpanzees and bonobos form a single genus: *Pan.* The human lineage diverged from the *Pan* ancestor about 5.5 million years ago. Some scientists feel that humans, chimpanzees, and bonobos are close enough to form a single genus: *Homo.* Since bonobos and chimpanzees split from each other after they split from us, about 2.5 million years ago, both are equally close to us. The gorilla diverged earlier, hence it is more distant from us, as is the only Asian great ape, the orangutan.

that the bonobo "may approach more closely to the common ancestor of chimpanzees and man than does any living chimpanzee."

Adaptation to life in the trees is obvious from the way bonobos use their bodies, which by human standards is rather unusual. Their feet serve as hands. They grasp things with their feet, gesture with them during communication, and clap them together to attract attention. Apes are sometimes called "quadrupedal" (four-footed), but bonobos might be better called "quadrumanual" (four-handed). They are more gymnastic than any other ape, jumping, brachiating, and flipping around in the trees with unbelievable agility. They walk along a suspended rope on two legs as if on terra firma. These acrobatic talents are practical for apes who have never been forced, not even partially, out of the forest, and who have hence never had to compromise their tree-dwelling ways. That bonobos are more arboreal than chimps is clear from a comparison of reactions to seeing scientists for the first time in the wild: Chimps drop straight out of the tree they are in and flee by running over the ground. Bonobos, on the other hand, flee through the tree canopy and only descend to the forest floor once they're far away.

I expect the debate about which ape most resembles the last common ancestor to go back and forth for a while, but for the moment, let us just say that chimps and bonobos are equally relevant for human evolution. The gorilla is set apart from both them and us by its enormous sexual dimorphism—the size difference between males and females—and the social system that goes with that: a single male monopolizing a harem of females. For simplicity's sake, I will mention gorillas only occasionally while we explore the similarities and differences among bonobos, chimps, and ourselves.

We do not wait around to see what happens between Azizi and Paul Donn. Undoubtedly, the two will make contact, but it could take hours, even days. The keepers realize that it will change Azizi's attitude forever: she will never be the same dependent little gorilla that Gale fed with a bottle and carried on his back until she grew too heavy. Her new destiny will be to live in this group, be around a big male of her own kind, and perhaps raise offspring.

We walk by the bonobos, where Loretta greets me with shrill hoots. Even though my research time at this zoo was almost twenty years ago, she still knows me: recognition is permanent. I cannot imagine ever forgetting a face that I saw every day over a period of time, so why should it be different for Loretta? And her hoots are distinctive. Bonobo calls are unmistakable: The easiest way to tell chimps and bonobos apart is just by listening to them. The low huu-huu sound of the chimpanzee is absent in the bonobo. Bonobo voices are so high pitched (more like hee-hee) that when the Hellabrunn Zoo in Munich received its first bonobos, the director almost sent them back. He hadn't yet looked under the cloth covering the crates from Bolobo and couldn't believe that the sounds he heard came from apes.

Loretta presents her ballooning genitals, staring at me upside-down though her legs, invitingly waving an arm. I wave back while asking Mike about one of the males who's not out. Mike takes me to the night quarters. The male sits indoors with a young female to keep him company. The female is clearly annoyed each time Mike turns to talk to me. What is this stranger doing here, and why is Mike not giving her his full attention? She tries to grab me through the bars. The male keeps his distance, but presents his back, then his belly to Mike for touching, while showing—as any male bonobo would under such circumstances—an impressive erection. For both male and female bonobos, there is no dividing line between sexuality and affection.

This male needs to be kept out of the bonobo group due to his low rank. Even though he's fully grown, he's unable to defend himself against an entire group of females. Female hostility against males is a growing problem among bonobos in zoos. In the past, the zoo community made a fundamental mistake by moving male bonobos around. If they had to send apes to another zoo for breeding, they always chose males. While this works well for most animals, it is a disaster for male bonobos. In nature, bonobo females do all the migrating, leaving their home group at puberty. Males stay put, enjoying the company and protection of their mothers. Males with influential mothers rise in the hierarchy and are tolerated around food. Zoos have learned the hard way that they need to follow this example. Unfortunately, the male in question had been brought in from the outside. Being real mama's boys, males fare best in the group in which they were born.

So aggression is by no means absent among bonobos. When females attack, things get ugly. From a screaming ball of arms and legs, it is invariably the male who emerges with injuries. Even though bonobos are usually great conciliators, they have these skills for a good reason: they are not above fighting. The bonobo provides a compelling example of social harmony precisely because underlying tensions remain visible. This paradox applies to us as well. In the same way that the ultimate test of a ship is how it holds up in a storm, we only fully trust a relation if it has survived occasional conflict.

After witnessing a few more sexual encounters among the bonobos, Mike cannot resist bringing up the recent claim of a local scientist that the zoo's bonobos rarely have sex, perhaps only a couple of times per year. Could it be that bonobos do not deserve their sexy reputation? Standing back outside among the public, we joke that since today we have counted six sexual encounters in just two hours, this must mean that we have gathered about two years' worth of observations. For a moment, I forget that Mike and Gale have their uniforms on, meaning that everyone around us is listening in. A bit too loudly, I boast about my earlier study, "When I was here, I had seven hundred sexual encounters in a single winter." A man standing next to us grabs his young daughter by the arm and rushes away.

Sometimes bonobo sex is subtle. One young female tries to walk past an even younger male who is blocking her path on a branch. The male fails to move out of the way—perhaps he is afraid of falling—whereupon the female places her teeth on his hand, which holds on to the branch, making matters worse. But instead of using force, she then turns around and rubs her clitoris against his arm. They are both immature, but this is the bonobo way of resolving conflict, a tactic that begins early in life. Following this contact, the female calmly climbs over the male and continues along the branch.

Upon returning home from San Diego, I am struck by the contrast with chimpanzees. I work with about forty chimps outdoors at the Field Station of the Yerkes National Primate Research Center, near Atlanta. I have known these apes for a long time, and I see them as distinct personalities. They know me equally well and pay me the one compliment every researcher craves, which is that they treat me like furniture. I walk up to the fence to say hello to Tara, the three-year-old little daughter of Rita, who sits high up in a climbing structure. Rita briefly glances down at us, then continues grooming her own mother, Tara's grandmother. If a stranger had merely walked by, Rita, who is extremely protective, would have immediately jumped down to pick up her daughter. I feel honored by her lack of interest.

I notice a deep, fresh gash on the upper lip of Socko, the second ranking male. Only one other male can have done this: Bjorn, the alpha male. Bjorn is smaller than Socko, but he's extremely smart, high-strung, and mean. He keeps the other apes under control by fighting dirty. This is the conclusion we have come to over the years, from seeing Bjorn's combat technique and the scars he leaves on his victims in unusual places, such as the belly or scrotum. Socko—a large, lumbering "ham"—cannot compete, so he must live under this little dictator. But luckily for Socko, his younger brother, who is in the midst of his final growth spurt, is eager to hang out with him. This is bound to create trouble for Bjorn very soon.

Here at Yerkes, we're in the thick of male power politics, the never-ending saga of chimpanzee society. Ultimately these battles are about females, which means that the fundamental difference between our two closest relatives is that one resolves sexual issues with power, while the other resolves power issues with sex.

## A VENEER OF CIVILIZATION

As I opened the newspaper on a plane ride from Chicago to Charleston, South Carolina, the first thing that caught my eye was the headline "Lili to Hit Charleston." This was disconcerting as Lili was a major hurricane, and the devastation of Hugo the year before was still fresh in everybody's memory. Lili swung away from Charleston, however, and the storm I ended up in was merely academic.

The conference I attended was about world peace and peaceful human relations. I went there to present my work on primate conflict resolution. It is always fun to speculate why people are drawn to certain fields, but peace research attracts its share of hotheads. At the meeting, two prominent peace advocates got into a shouting match, something having to do with the first referring to studies of Eskimos and the second accusing him of colonialist if not racist attitudes, since these people ought to be called Inuit. According to the book *Never in Anger,* Inuit go to extreme lengths to avoid exchanges that even remotely resemble hostility. Anybody who raises his or her voice risks being ostracized, which is a life-threatening punishment in their environment.

Some of us at the meeting surely would have been put out on the ice. Being Westerners, nonconfrontation was not in our script. I could already see another newspaper headline, something to the effect of "Peace Conference Ends in Fistfight." It is the only academic event at which I have seen fully grown men walk out of the room slamming doors like little children. Amid all this grandstanding, some dared wonder, with deep scholarly frowns on their faces, if human and ape behavior were truly comparable.

On the other hand, I have attended many meetings of the Aggression

Club, a group of academics in the Netherlands, which were always civil and calm. Though only a graduate student at the time, I was allowed to join psychiatrists, criminologists, psychologists, and ethologists who regularly came together to discuss aggression and violence. In those days, evolutionary views invariably revolved around aggressiveness, as if our species had no other behavioral tendencies to speak of. It was like a discussion about pit bulls in which the main topic is always how dangerous these dogs are. What makes humans different from pit bulls, however, is that we have not been selectively bred to fight. Our jaw pressure is miserable, and our brains certainly would not need to be the size they are if the only trait that mattered was the killing of others. But in the postwar era, human aggressiveness was front and center of every debate.

With its gas chambers, mass executions, and willful destruction, World War II was human behavior at its worst. Moreover, when the Western world took stock after the dust had settled, it was impossible to ignore the savagery that had been committed in the heart of Europe by otherwise civilized people. Comparisons with animals were ubiquitous. Animals lack inhibitions, the argument went. They lack culture, so it must have been something animal-like, something in our genetic makeup that had burst through the veneer of civilization and pushed human decency aside.

This "veneer theory," as I call it, became a dominant theme in the postwar discussion. Deep down, we humans are violent and amoral. A stream of popular books explored this issue by proposing that we have an uncontainable aggressive drive that seeks an outlet in warfare, violence, and even sports. Another theory was that our aggressiveness is novel, that we are the only primates that kill their own kind. Our species never had the time to evolve the appropriate inhibitions. As a result, we don't have our fighting instinct under control as much as "professional predators" like wolves or lions. We're stuck with a violent temper that we're ill-equipped to master.

It is not hard to see the beginnings here of a rationalization of human violence in general and the Holocaust in particular, and it certainly did not help that the leading voice of the time spoke German. Konrad Lorenz, a world-renowned Austrian expert on fish and geese, was the great defender of the idea that aggression is in our genes. Killing became humanity's "mark of Cain."

Across the Atlantic, a similar view was promoted by Robert Ardrey, an American journalist inspired by speculations that *Australopithecus* must have been a carnivore who swallowed his prey alive, dismembering them limb from limb, slaking his thirst with warm blood. Drawn from studies of a few skull bones, this was an imaginative conclusion, but Ardrey based his killer ape myth on it. In *African Genesis*, he painted our ancestor as a mentally deranged predator upsetting the precarious balance of nature. In Ardrey's demagogic prose, "We were born of risen apes, not fallen angels, and the apes were armed killers besides. And so what shall we wonder at? Our murders and massacres and missiles, and our irreconcilable regiments?"

It's hard to believe, but the next wave of pop biology managed to go beyond this. At the same time that Ronald Reagan and Margaret Thatcher preached that greed was good for society, good for the economy, and certainly good for those with anything to be greedy about, biologists published books in support of these views. Richard Dawkins's *The Selfish Gene* taught us that since evolution helps those who help themselves, selfishness should be looked at as a driving force for change rather than a flaw that drags us down. We may be nasty apes, but it makes sense that we are, and the world is a better place for it.

A tiny problem—pointed out to no avail by nitpickers—was the misleading language of this genre of books. Genes that produce successful traits spread in the population and hence promote themselves. But to call this "selfish" is nothing but a metaphor. A snowball rolling down the hill gathering more snow also promotes itself, but we generally don't call snowballs selfish. Taken to its extreme, the everything-is-selfishness position leads to a nightmarish world. Having an excellent nose for shock value, these authors haul us to a Hobbesian arena in which it's every man for himself, where people show generosity only to trick others. Love is unheard of, sympathy is absent, and goodness a mere illusion. The best-known quote of those days, from biologist Michael Ghiselin, says it all, "Scratch an altruist, and watch a hypocrite bleed."

We should be happy that this dark, forbidding place is pure fantasy, that it differs radically from the actual world in which we laugh, cry, make love, and fawn over babies. Authors of this fiction realize this and sometimes confide that the human condition is not as bad as they make it sound. *The Selfish Gene* is a good example. Having argued that our genes know what is best for us, that they program every little wheel of the human survival machine, Dawkins waits until the very last sentence of his book to reassure us that, in fact, we are welcome to chuck all those genes out the window: "We, alone on earth, can rebel against the tyranny of the selfish replicators."

And so the end of the twentieth century emphasized our need to rise above nature. This view was advertised as Darwinian, even though Darwin had nothing to do with it. Darwin believed, as I do, that our humaneness is grounded in the social instincts that we share with other animals. This is obviously a more optimistic view than the one proclaiming that we "alone on earth" can overcome our basic instincts. In the latter view, human decency is no more than a thin crust—something we invented rather than inherited. And each time we do anything less than honorable, veneer theorists will remind us of the dreadful core underneath: "See, there's human nature!"

## OUR DEMONIC FACE

The opening scene of Stanley Kubrick's *2001: A Space Odyssey* captured in one dazzling image the idea that violence is good. After a fight breaks out among hominids, in which one bludgeons another with a zebra femur, the weapon is flung triumphantly into the air where millennia later it turns into an orbiting spacecraft.

Equating aggression with progress underlies the so-called "Out of Africa" hypothesis, which posits that we've gotten to where we are today through genocide. When bands of *Homo sapiens* migrated out of Africa, they advanced into Eurasia by murdering all other bipedal apes they ran into, including the species most similar to them, the Neanderthals. Our bloodthirstiness is the centerpiece of books with titles like *Man the Hunter*, *Demonic Males*, *The Imperial Animal*, and *The Dark Side of Man*, which mostly feature the chimpanzee as our ancestral model—the *male* chimpanzee. Like the bombshells in early James Bond movies, females are what males fight over, but apart from being mates and mothers, they barely enter the story. Males do all the deciding and fighting—and, by implication, most of the evolving.

But even though the chimpanzee has come to represent the demonic face of our Janus head, it has not always been this way. Around the time that Lorenz and Ardrey were highlighting our "mark of Cain," wild chimps seemed to be doing little else than moving lazily from tree to tree collecting fruit. Adversaries of the killer ape view—and they were plentiful—used this information to their advantage. They freely quoted Jane Goodall, who in 1960 had started her work at Gombe Stream in Tanzania. At the time, Goodall still presented chimps as the noble savages of French philosopher Jean-Jacques Rousseau: self-sufficient loners with no need to connect or to compete with one another. Chimps in the jungle traveled alone or in small "parties" that changed composition all the time. The only permanent bonds were those between mother and dependent offspring. No wonder people thought of apes as living in Eden.

The first correction of these impressions came in the 1970s from Japanese scientists who studied chimpanzees a little south of Gombe, in the Mahale Mountains. They had serious doubts about the "individualistic" bias of American and European researchers. How could an animal so close to us have no society to speak of? They found that even though chimps hang out with different companions every day, they're all members of a single community that is separate from other communities.

The second correction was to the reputation of wild chimpanzees as peaceful, which some anthropologists used as an argument against an innately aggressive human nature. Two reality checks appeared. First, we heard that chimps hunt monkeys, smash in their skulls, and eat them alive. This made them carnivores. Then, in 1979, the posh pages of *National Geographic* reported that these apes also kill each other, sometimes consuming their victims. This made them both murderers and cannibals. The report showed sketches of male chimps stalking unsuspecting enemies across the border of their territory, surrounding them, and brutally beating them to death. At first, this news trickled in from only a few sources, but soon the trickle turned into a steady stream that was impossible to ignore.

The picture became indistinguishable from that of the killer ape. We now knew that chimps hunt to kill and live in communities that are at war with one another. In a later book, Goodall relates how she broke this news to a group of academics, some of whom held out hope of eliminating human aggression through education and improved television programming. Her message that we are not the only aggressive primates was unwelcome: Shocked colleagues begged her to downplay the evidence or even not to publish it. Others felt the camp at Gombe, where investigators dispensed bananas—an unnatural, highly nutritious food—had fostered pathological levels of aggression. Competition at the campsite was indeed well documented, but the most serious fighting had taken place far away from there. Goodall resisted her critics: "Certainly, I felt strongly it was better to face up to the facts, however unsettling, than to live in a state of denial," she said.

The banana critique failed to stick: warfare is now also known from African sites without any provision of extra food. The simple truth is that brutal violence is part of the chimp's natural makeup. They don't *need* to show it—some chimpanzee communities indeed seem rather peaceful—

but they can and often do. This strengthens the killer ape theory in one way, but it also undermines it. Lorenz and Ardrey claimed that humans stood alone in their use of lethal force, whereas observations of not only chimps but also hyenas, lions, langurs, and a long list of other animals have since made it clear that killing one's own kind, while infrequent, is in fact widespread. Ed Wilson, the sociobiologist, concluded that by the time a particular animal is observed for more than a thousand hours, scientists will see mortal combat. Here, he spoke as an expert on ants, a group of insects that raids and kills on a massive scale. In Wilson's words, "Alongside ants, which conduct assassinations, skirmishes, and pitched battles as routine business, men are all but tranquilized pacifists."

With the discovery of the chimp's dark side—"paradise lost"—Rousseau went out the door and Hobbes came in. The violence of apes surely meant that we are programmed to be ruthless. Combined with evolutionary biologists' claim that we are genetically selfish, everything clicked into place. There was now a coherent, irrefutable view of humanity. Look at the chimpanzee, the argument went, and you will see what kind of monsters we truly are.

Chimps thus fortified the idea of a nasty human nature despite the fact that, without too much trouble, they could equally well have countered it. After all, chimp violence is far from an everyday occurrence: it took scientists decades to reveal it. Unhappy with the one-sided impact of her discoveries, Goodall herself made valiant efforts to illuminate the gentler side of chimpanzees, their compassion even, but to no avail. Science had made up its mind: once a murderer, always a murderer.

Chimps may be violent, but at the same time their communities have powerful checks and balances. This became clear to me one day at the Arnhem Zoo in the Netherlands when we stood in nail-biting anticipation at the edge of a moat surrounding a forested island. We were worried about a tiny newborn chimpanzee known as Roosje, Dutch for "little rose." Roosje had been adopted by Kuif. Having no milk of her own, Kuif had been trained to give Roosje the bottle. The plan had worked beyond our wildest dreams. A minor feat for an ape was a giant success for us, or so we felt. But now we were reintroducing the mother with her new daughter to the world's largest zoo colony of chimpanzees, one that included four dangerous adult males. To intimidate their rivals, males tend to charge about with all their hair on end so that they look large and threatening. Unfortunately, this was the state that Nikkie, the colony's fearless leader, was in.

Male chimps have a ferocious temper and are so strong that they can overpower humans easily; when angry they are beyond our control. So Roosje's fate was in the apes' hands. In the morning, we had paraded Kuif past all of the night cages to gauge the group's reaction. They all knew Kuif, but Roosje was new. When Kuif walked by the male cage, something caught my attention. Nikkie was grabbing underneath her through the bars, causing her to jump away with a sharp yelp. His target seemed to be the spot where Roosje was clinging to Kuif's belly. Since only Nikkie acted this way, I decided to do the group introduction in stages, putting Nikkie at a disadvantage by releasing him last. The main thing to avoid was letting Kuif find herself alone with him. I counted on her protectors in the group.

In the wild, chimps occasionally kill infants of their own species. Some biologists' theories about these acts of infanticide assume that males compete over the fertilization of females. This would explain their constant jockeying for position as well as the killing of infants that are not their own. Nikkie may have seen Roosje as an outside infant, one that could not possibly have been fathered by him. This was hardly reassuring as it meant that we might witness one of those gruesome scenes reported from the field. Roosje might be torn to shreds. Since I had been holding her for weeks, helping Kuif to feed her, and feeding her myself, I was far from the dispassionate observer I normally like to be.

Once on the island, most colony members greeted Kuif with an embrace, stealing glances at the baby. Everyone seemed to be keeping a nervous eye on the door behind which Nikkie sat waiting. Some youngsters hung around the door, kicking it, waiting to see what would happen. All this time, the two oldest males stayed close to Kuif and were extremely friendly to her.

After about an hour, we let Nikkie onto the island. The two males left Kuif and positioned themselves between her and the approaching Nikkie, their arms draped around each other's shoulders. This was a sight to behold, given that they had been archenemies for years. Here they were, standing united against the young leader, perhaps fearing the same thing we feared. Nikkie, who had all his hair up, approached in a most intimidating manner but broke down when he saw that the other two were in no mood to budge. Kuif's defense team must have looked incredibly determined, staring down the boss like that, because Nikkie fled. I couldn't see their faces, but apes read as much in each other's eyes as we do. Nikkie later approached Kuif under the watchful eyes of the other two males. He was nothing but gentle. His intentions will forever remain a mystery, but we sighed an enormous sigh of relief and I hugged the caretaker with whom I had run Kuif's training.

Chimps live under a cloud of potential violence, and infanticide is a leading cause of death both at zoos and in the wild. But in the end, when debating how aggressive *we* are as a species, chimpanzee behavior is only one piece of the puzzle. The behavior of our immediate ancestors would be more relevant. Unfortunately, there exist enormous holes in our knowledge of them, especially if we try to look back further than ten thousand years. There is no firm evidence that we have always been as violent as in the past few millennia. From an evolutionary standpoint, a few thousand years is nothing.

For millions of years before, our ancestors may have led an easygoing existence in small groups of hunter-gatherers who had little to fight over, given how thinly populated the world was then. This would by no means have kept them from conquering the globe. It's often thought that survival of the fittest means wiping out the unfit. But one can also win the evolutionary race by having a superior immune system or by being better at finding food. Direct combat is rarely the way one species replaces another. Thus, instead of annihilating the Neanderthals, we may simply have been more resistant to the cold or have been better hunters.

It's entirely possible that successful hominids "absorbed" less successful ones through crossbreeding, and therefore whether or not Neanderthal genes survive in you and me is an open question. If people joke that someone looks like a Neanderthal, they should think twice. In a Moscow laboratory I once saw a remarkable reconstruction of a Neanderthal face based on a skull. The scientists confided that they had never dared publicize the bust due to its uncanny resemblance to one of their country's leading politicians, who might not have appreciated the comparison.

## THE APE IN THE CLOSET

Would scratching a bonobo expose a hypocrite?

We can be pretty sure that the notorious synopsis of veneer theory concerned people only. No one would suggest that animals are pulling the wool over each other's eyes. This is why apes are crucial to the debate about the human condition. If they turn out to be better than brutes—even if only occasionally—the notion of niceness as a human invention begins to wobble. And if true pillars of morality, such as sympathy and intentional altruism, can be found in other animals, we will be forced to reject veneer theory altogether. Darwin was aware of these implications when he observed that "many animals certainly sympathize with each other's distress or danger."

Of course they do. It's not unusual for apes to care for an injured companion, slowing down if another lags behind, cleaning another's wounds, or carrying fruit down from a tree to an elder who has lost her climbing abilities. One field report tells of an adult male chimp who adopted an orphan, carrying the sickly infant during travel, shielding it from danger, saving its life even though the two were presumed unrelated. In the 1920s, the American ape expert Robert Yerkes was so struck by the concern shown by one young chimp, Prince Chim, for his terminally ill companion, Panzee, that he admitted, "If I were to tell of his altruistic and obviously sympathetic behavior towards Panzee I should be suspected of idealizing an ape." Yerkes's admiration for Prince Chim's sensitivity is telling, given that he probably knew more ape personalities than anyone else in the history of primatology. He paid tribute to this kind little anthropoid in a book titled *Almost Human*, in which he expressed doubt that Prince Chim was a regular chimp. A postmortem inspection has since revealed that he was indeed no chimp, but a bonobo. Yerkes did not know this—the bonobo wasn't recognized as a species until years later.

The first study to compare the behavior of bonobos and chimpanzees was carried out in the 1930s at the Hellabrunn Zoo. Eduard Tratz and Heinz Heck published their findings in 1954. Three bonobos had been terrified by the city's bombardment one night during the war and had died of heart failure. The fact that all of the zoo's bonobos died of fright but that none of the zoo's chimps suffered the same fate attests to the bonobo's sensitivity. Tratz and Heck made a long list of the differences between bonobos and chimps, which included references to the bonobo's relative peacefulness, sexual behavior, and sensuality. Aggression is certainly not absent in the bonobo, but the treatment to which chimps occasionally subject one another, including biting and full-force hitting, is rare in bonobos. The hair of a male chimp stands on end at the slightest provocation. He will pick up a branch and challenge anyone perceived as weaker. Chimps are very much into status. By bonobo standards, the chimpanzee is a wild and untamed beast, or as Tratz and Heck put it, "The bonobo is an extraordinarily sensitive, gentle creature, far removed from the demoniacal *Urkraft* [primitive force] of the adult chimpanzee."

If this was known in 1954, one might ask, Why was the bonobo absent from the debates on human aggression, and why is it still not more widely known? Well, that study was published in German, and the time when English-speaking scientists read anything other than English is long past.

Also, this study included only a few young apes in captivity, a tiny scientific sampling, so it may not have sounded too convincing. Bonobo field research, which got underway relatively late, still lags decades behind studies on the other great apes. Another reason is cultural: the bonobo's eroticism was a topic few authors wanted to touch. This remains the case today. In the 1990s, a British camera crew traveled to the remote jungles of Africa to film bonobos only to stop their cameras each time an "embarrassing" scene appeared in the viewfinder. When a Japanese scientist assisting the crew asked why they weren't documenting any sex, he was told "our viewers wouldn't be interested."

But far more important than all of this is the fact that bonobos fail to fit established notions about human nature. Believe me, if studies had found that they massacre one another, everyone would know about bonobos. Their peacefulness is the real problem. I sometimes try to imagine what would have happened if we'd known the bonobo first and the chimpanzee only later or not at all. The discussion about human evolution might not revolve as much around violence, warfare, and male dominance, but rather around sexuality, empathy, caring, and cooperation. What a different intellectual landscape we would occupy!

It's only with the appearance of another of our cousins that the stranglehold of the killer ape theory began to loosen. Bonobos act as if they have never heard of the idea. Among bonobos, there's no deadly warfare, little hunting, no male dominance, and enormous amounts of sex. If the chimpanzee is our demonic face, the bonobo must be our angelic one. Bonobos make love, not war. They're the hippies of the primate world. Science had more trouble with them than a 1960s family had with its long-haired, pot-smoking black sheep who wanted to move back home. They turned off the lights, hid under the table, and hoped that the uninvited guest would go away.

The bonobo is clearly an ape for our time. Attitudes have changed dramatically since Margaret Thatcher postulated her strident individualism. "There is no such thing as society," she proclaimed. "There are individual men and women, and there are families." Thatcher's comment may have been inspired by the evolutionary views of her day, or perhaps it was the other way around. Either way, twenty years later, when huge corporate scandals have delivered the final pinprick to an inflated stock balloon, pure individualism does not sound so hot anymore. In the post-Enron era the public has begun to realize again—as if it had not always known—that unmitigated capitalism rarely brings out the best in people. Reagan and Thatcher's "gospel of greed" went sour. Even Federal Reserve Chairman Alan Greenspan, a prophet of capitalism, hinted that it might be good to put the brakes on, explaining to a U.S. Senate committee in 2002, "It is not that humans have become any more greedy than in generations past. It is that the avenues to express greed have grown so enormously."

Anyone following evolutionary biology must have noticed a parallel change of heart. Suddenly books appeared with titles like *Unto Others, Evolutionary Origins of Morality, The Tending Instinct, The Cooperative Gene,* and my own *Good Natured.* There was less talk of aggression and competition and more of connectedness, of how societies hang together, of the origins of caring and commitment. What was stressed was the *enlightened* self-interest of the individual within a larger whole. Whenever interests overlap, competition will be constrained by the greater good.

Along with other economic gurus of the era, Klaus Schwab declared that it was time for business "governed not only by rules but by values" while evolutionary biologists began to insist that "the rational pursuit of self-interest is sometimes an inferior strategy." Perhaps both developments derived from broader swings in public attitude. Having rebuilt economies destroyed by war and having reached a level of prosperity unimaginable not too long ago, the industrialized world may finally be ready to focus on the social domain. We need to decide whether we are like Robinson Crusoes sitting on separate little islands, as Thatcher seemed to imagine, or members of complexly interwoven societies in which we care about each other and from which we derive our reason for being.

Being more in agreement with the second possibility than the first, Darwin felt that people are born to become moral and that animal behavior supports this idea. He relates how a dog he knew would never walk by a basket where a sick friend lay, a cat, without giving her a few licks with his tongue. This, Darwin said, was a sure sign of the dog's kind feelings. Darwin also tells the story of a zoo keeper with a wound at the nape of his neck. It was inflicted by a fierce baboon while the keeper was cleaning his cage. The baboon lived with a little South American monkey. Scared to death of his companion, the monkey was a warm friend to the keeper and in fact saved his life by distracting the baboon with bites and screams during his assault. The small monkey thus risked his own life, showing that friendship translates into altruism. Darwin felt it was the same for people.

This was before we knew of the bonobo and before the latest findings of neuroscience. By placing people in brain scanners and asking them to resolve moral dilemmas, experts have discovered that such dilemmas activate ancient emotional centers deeply embedded in the brain. Instead of being a surface phenomenon in our expanded neocortex, moral decision-making apparently taps into millions of years of social evolution.

Perhaps this sounds obvious, but it's monumentally at odds with the view of morality as a cultural or religious veneer. I've often wondered how a position so plainly wrong could have been advocated for so many years. Why were altruists seen as hypocrites, why were emotions left out of the debate, and why did a book boldly titled *The Moral Animal* deny that morality comes naturally to us? The answer is that evolutionary authors were making the "Beethoven error." By this, I mean the assumption that process and product need to resemble each other.

Listening to Ludwig van Beethoven's perfectly structured music, one would never guess what his poorly heated apartment looked like. Visitors complained that the composer lived in the dirtiest, smelliest, most disorderly place imaginable, strewn with wasting food, unemptied chamber pots, and filthy clothing, his two pianos buried under dust and papers. The maestro himself looked so slovenly that he was once arrested as a vagrant. No one asks how Beethoven could have created his intricate sonatas and noble piano concertos in such a pigsty. We all know that wonderful things can originate under atrocious circumstances, that process and product are separate things, which is why the enjoyment of a good restaurant is rarely enhanced by a visit to its kitchen.

Yet confusion between the two has led some to believe that since natural selection is a cruel, pitiless process of elimination it must produce cruel and pitiless creatures. A nasty process must produce nasty behavior, or so the thinking went. But nature's pressure cooker has created both fish that lunge at anything that moves (including their own offspring) and pilot whales, which are so attached to each other that they beach themselves together if one of them gets disoriented. Natural selection favors organisms that survive and reproduce, pure and simple. How they accomplish this is left open. Any organism that can do better by becoming either more or less aggressive than the rest, more or less cooperative, or more or less caring will spread its genes. The process does not specify the road to success any more than the inside of a Viennese apartment tells us what kind of music will be floating out its window.

## ANALYZING APES

Every late afternoon at the Arnhem Zoo, the keepers and I would call Kuif out of the colony for the daily bottle-feeding of Roosje. Before she would come in carrying her adopted child, however, there always was a strange ritual.

We're used to apes greeting each other, which they do after long absences, by either kissing and embracing (chimps) or some sexual frottage (bonobos). Kuif was the first ape I have ever seen do good-bye (we call it "saying good-bye," but obviously an ape can only "do good-bye"). Before she would enter the building, Kuif would approach Mama, the respected alpha female of the group and her best friend, to give her a kiss. After this, she would seek out Yeroen, the oldest male, to do the same. Even if Yeroen was asleep at the far end of the island, or deep into a grooming session with one of his buddies, Kuif would make a grand detour to catch him. It reminded me of how we're not to leave a party without saying farewell to the hosts.

For greeting, all that's needed is delight at the sight of a familiar individual. Many social animals display this response. Saying good-bye is more complex, as it requires a glance into the future: the realization that you will not see someone for a while. I noticed another such glance one day when a female chimp collected all the straw from her night cage, deliberately scooping up every last bit until her arms were overflowing, and carried it all outside to the island. No chimp ever walks around with straw, so this caught our attention. It was November, and the days were getting cold. Apparently, this female had decided she was going to stay warm outside. At the moment she gathered the straw she could not feel the cold—she was in a heated building—so she must have extrapolated from the previous day to the next. She spent the whole day in her straw nest, which she couldn't leave since everyone else was waiting to steal from it.

This is the sort of intelligence that draws many of us to the study of apes. Not just their aggressive or sexual behavior, much of which they share with other animals, but the surprising amount of insight and finesse they put into everything they do. Since much of this intelligence is hard to pinpoint, studies on captive apes are absolutely essential. In the same way that no one would try to measure a child's intelligence by watching him or her run around in the school yard, the study of ape cognition demands a hands-on approach. One needs to be able to present the apes with problems to see how they solve them. Another advantage of captive apes under enlightened conditions (meaning spacious outdoor areas and a naturalistic group size) is that one can watch their behavior much more closely than is possible in the field, where at critical moments they tend to disappear in the undergrowth.

My favorite office (I have several) has a large window overlooking the chimps at the Yerkes Field Station, allowing me to see everything that goes on. They can't hide from me (nor I from them, as is clear each time I try to eat my lunch unnoticed). Simple observation is the reason that power politics, reconciliations after fights, and tool-use were first discovered in captive apes and were confirmed only later in the wild. Typically, we conduct observations with binoculars and a keyboard attached to a computer into which we type every social event we witness. We have a long list of codes for play, sex, aggression, grooming, nursing, and myriad subtle distinctions within each category, and we enter data on a continuous basis in the format of "who does what to whom." If events become too complicated, such as when a massive fight breaks out, we film it or, like a sportscaster, we narrate the events into a tape recorder. This way, we gather literally hundreds of thousands of observations, then program a computer to sort through the data. Despite the pleasure we take in our work, primate research has its tedious side.

When we want to present the apes with a problem, we call them out of their group into a small building. Since we cannot force them to participate, we depend on their willingness. Not only do all the apes know their names, they also know each other's names, so that one can ask individual A to fetch B. The trick is, of course, to make the experiment a pleasant experience. Computers with joysticks really attract them. My assistant only needs to show the cart with equipment and the volunteers will be lining up. As with children, the immediate feedback of a computer gives apes a thrill.

In one experiment, Lisa Parr presented Yerkes chimpanzees with hundreds of pictures that I had taken at the Arnhem Zoo. With an ocean between these chimps, we could be sure they had never seen those faces before.

One face would appear on the computer screen and then two more faces, one of which matched the first. The chimpanzee would be given a sip of juice for moving the cursor to the matching face. Face recognition had been tested before, but apes were not particularly good at it. Previous experiments, however, had used human faces on the assumption that these are easy to tell apart. Not for chimpanzees. They were far better with the faces of other chimps. Lisa showed that they see similarities not only between different pictures of the same face, but also between pictures of mother and offspring. In the same way that if I went through your family album, I could probably tell your relatives from your in-laws, chimps recognize the marks of kinship. They seem just as sensitive to their faces as we are to ours.

Another study asked if chimps can deliberately point things out to others. The earlier story of Kanzi and Tamuli suggests they can, but the claim remains controversial. Some scientists focus on pointing with the hand or index finger, which is the way we point. However, I see no reason to take such a limited view. Nikkie once communicated with me using a much more subtle technique. He had gotten used to my throwing wild berries across the moat. One day while I was taking data I had totally forgotten about the berries, which hung on a row of tall bushes behind me. Nikkie had not. He sat down right in front of me, locked his red-brown eyes on mine, and—once he had my attention—abruptly jerked his head and eyes away from mine to fixate on a point over my left shoulder. He then looked back at me and repeated the move. I may be dense compared to an ape, but the second time I followed his gaze, spotting the berries. Nikkie had indicated what he wanted without a single sound or hand gesture. There is obviously no point to such "pointing" unless you understand that another hasn't seen what you have seen, which means realizing that not everyone has the same information.

A telling experiment on ape pointing was conducted by Charles Menzel at the same Language Research Center that is home to Kanzi. Charlie let a female chimpanzee named Panzee watch while he hid food in a forested area near Panzee's cage. Panzee followed the food-hiding from behind the bars. Since she could not go where Charlie was, she would need human help to get the food. Charlie would dig a small hole in the ground and hide a bag of M&M's or place a candy bar in some bushes. Sometimes he would do this after all people had left for the day. This meant that Panzee could not communicate with anybody about what she knew until the next day. When caretakers arrived in the morning, they didn't know about the experiment. Panzee first had to get their attention and then provide information to someone who did not know what she knew and who at first had no clue what she was "talking" about.

During a live demonstration of Panzee's skills, Charlie noted as an aside to me that caretakers generally have a higher opinion of apes' mental abilities than the philosophers and psychologists who write on the topic, few of whom have interacted with these animals on a daily basis. It was essential for the experiment, he explained, that Panzee dealt with people who took her seriously. All those recruited by Panzee said they were at first surprised by her behavior, but soon they understood what she was trying to get them to do. Following her pointing, beckoning, panting, and calling, they had no trouble finding the item hidden in the forest. Without her instructions, they would not have known where to look. Panzee never pointed in the wrong direction or to any location used on earlier occasions. The result was communication about a past event, present in the ape's memory, to people who knew nothing about it and were unable to give her any clues.

I offer these examples to make the point that there is excellent research on apes to rely upon when making statements about their sense of the past, sense of the future, face recognition, and social behavior in general. Even though in this book I favor lively examples, trying to put a face on what we know about our nearest relations, there exists an entire body of academic literature to back up most of my claims. Not all of them, mind you, which explains why disagreements persist and why there is no end in sight for my line of work. A conference on the great apes might attract one or two hundred experts, but this is nothing compared to the typical meeting of psychologists or sociologists, which easily counts ten thousand. As a result, we are nowhere near the level of understanding of apes that many of us would like to be.

Most of my colleagues are field-workers. Whatever the advantages of research on captive apes, it can never replace the study of natural behavior. For every remarkable ability demonstrated in the laboratory, we want to know what it means for wild chimps and bonobos, what kind of benefits they gain from it. This relates to the evolutionary question, Why did the ability arise in the first place? For face recognition the benefits are pretty obvious, but what about a sense of the future? Field-workers have discovered that traveling chimpanzees sometimes gather grass stems and small twigs hours before arriving at a site where they will be fishing for ants or termites. The tools they need are picked up on the way at a place where they are abundant. It's entirely possible that apes plan their travel routes with this in mind.

What's perhaps most significant about this research is *not* what apes reveal about our instinctual side. With their slow development (they reach adulthood at around sixteen) and ample learning opportunities, apes are really not that much more instinctual than we are. Apes make lots of decisions in their lives, such as whether to threaten a newborn or defend it or whether to save a bird or abuse it. What we compare, therefore, are the ways in which humans and apes handle problems through a combination of natural tendencies, intelligence, and experience. It's impossible to extract from this mixture what is inborn and what is not.

Nevertheless, the comparison is instructive if only because it makes us step back and look into a mirror that shows a different side of ourselves than we are used to. You place your hand in a bonobo's and notice that your thumb is longer, you grab a bonobo's upper arm and have never felt muscles that hard, you pull at its lower lip and feel how much more lip it has than you, you look into its eyes and get a look back that is exactly as inquisitive as yours. All of this is revealing. My goal is to make the same comparisons with regard to social life and to show that there is not a single tendency that we do not share with these hairy characters we love to make fun of.

If people laugh at primates at the zoo, they do so, I suspect, precisely because they're unsettled by the mirror held up to them. Otherwise, why do strange-looking animals such as giraffes and kangaroos not cause similar hilarity? Primates arouse a certain nervousness because they show us ourselves in a brutally honest light, reminding us, in Desmond Morris's felicitous phrase, that we are mere "naked apes." It's this honest light that we seek, or ought to seek, and the beauty is that now that we know more about the bonobo, we can see ourselves reflected in two complementary mirrors.

TWO

# POWER

## Machiavelli in Our Blood

I put for a generall inclination of all mankind, a perpetuall and restlesse desire of Power after power, that ceaseth only in Death.

—THOMAS HOBBES

Egalitarianism is not simply the *absence* of a headman, but a positive insistence on the essential equality of all people and a refusal to bow to the authority of others.

—RICHARD LEE

Pedaling up one of the rare hills of my native Holland, I was bracing myself for the gruesome sight awaiting me at the Arnhem Zoo. In the early morning, I had received a call telling me that my favorite male chimpanzee, Luit, had been butchered by his own kind. Apes can inflict incredible damage with their powerful canine teeth. Most of the time, they are just trying to intimidate each other with what we call "bluff" displays, but occasionally the bluff is backed up by action. I had left the zoo the previous day worrying about Luit, but I was totally unprepared for what I found.

Normally proud and not particularly affectionate to people, Luit now wanted to be touched. He was sitting in a pool of blood, his head leaning against the bars of the night cage. When I gently stroked him, he let out the deepest sigh. Bonding at last, but at the saddest moment of my career as a primatologist. It was immediately obvious that Luit's condition was life-threatening. He still moved about but had lost enormous amounts of blood. He had deep puncture holes all over his body and had lost fingers and toes. We soon discovered that he was missing even more vital parts.

I have come to think of this moment in which Luit looked to me for comfort as an allegory of modern humanity: like violent apes, covered in our own blood, we long for reassurance. Despite our tendency to maim and kill, we want to hear that everything will be all right. At the time, however, I was focused only on trying to save Luit's life. As soon as the vet arrived, we tranquilized Luit and took him into surgery, where we sewed literally hundreds of stitches. It was during this desperate operation that we discovered Luit's testicles were gone. They had disappeared from the scrotal sac even though the holes in the skin seemed smaller than the testicles themselves, which the keepers had found lying in the straw on the cage floor.

"Squeezed out," the vet concluded impassively.

## TWO AGAINST ONE

Luit never came out of the anesthesia. He paid dearly for having stood up to two other males, frustrating them by his steep ascent. Those two had been plotting against him in order to take back the power they had lost. The shocking way they did so opened my eyes to how deadly seriously chimpanzees take their politics.

Two-against-one maneuvering is what lends chimpanzee power struggles both their richness and their danger. Coalitions are key. No male can rule by himself, at least not for long, because the group as a whole can overthrow anybody. Chimpanzees are so clever about banding together that a leader needs allies to fortify his position as well as the greater community's acceptance. Staying on top is a balancing act between forcefully asserting dominance, keeping supporters happy, and avoiding mass revolt. If this sounds familiar, it's because human politics works exactly the same.

Before Luit's death, the Arnhem colony was ruled jointly by Nikkie, a young upstart, and Yeroen, an over-the-hill conniver. Barely adult at seventeen, Nikkie was a brawny character with a dopey expression. He was very determined, but not the sharpest knife in the drawer. He was supported by Yeroen, who was physically not up to the task of being a leader anymore, yet who wielded enormous influence behind the scenes. Yeroen had a habit of watching disputes unfold from a distance, stepping in only when emotions were flaring to calmly support one side or the other, thus forcing everybody to pay attention to his decisions. Yeroen shrewdly exploited the rivalries among younger and stronger males.

Without going into the complex history of this group, it was clear that Yeroen hated Luit, who had wrested power from him years before. Luit had defeated Yeroen in a struggle that had taken three hot summer months of daily tensions involving the entire colony. The next year, Yeroen had gotten even by helping Nikkie dethrone Luit. Ever since, Nikkie had been the alpha male with Yeroen as his right-hand man. The two became inseparable. Luit was unafraid of either one of them alone. In one-on-one encounters in the night cages, Luit dominated every other male in the colony, taking away their food or chasing them around. No single one of them could possibly have kept him in his place.

This meant that Yeroen and Nikkie ruled as a team, and only as a team. They did so for four long years. But their coalition eventually began to unravel, and as is not uncommon among men, the divisive issue was sex. Being the kingmaker, Yeroen had enjoyed extraordinary sexual privileges. Nikkie would not let any other males get near the most attractive females, but for Yeroen he had always made an exception. This was part of the deal: Nikkie had the power, and Yeroen got a slice of the sexual pie. This happy arrangement ended only when Nikkie tried to renegotiate its terms. In the four years of his rule, he had grown increasingly self-confident. Had he forgotten who had helped him get to the top? When the young leader began to throw his weight around, interfering with the sexual adventures not only of other males but also of Yeroen himself, things got ugly.

Infighting within the ruling coalition went on for months, until one day Yeroen and Nikkie failed to reconcile after a spat. With Nikkie following him around, screaming and begging for their customary embrace, the old fox finally walked away without looking back. He'd had it. Luit filled the power vacuum overnight. The most magnificent chimpanzee male I have known, both in body and spirit, quickly grew in stature as the alpha male. Luit was popular with females, a mighty arbiter of disputes, protector of the downtrodden, and effective at disrupting bonding among rivals in the divide-and-rule tactic typical of both chimp and man. As soon as Luit saw other males together he would either join them or perform a charging display to disband them.

Nikkie and Yeroen both looked awfully depressed about their sudden loss of status. They seemed to have shrunk in size. But at times they seemed ready to resurrect their old coalition. That this actually happened in the night quarters, where Luit had no escape, was probably no accident. The horrible scene the keepers discovered told us that Nikkie and Yeroen had not only repaired their differences, but had acted together in highly coordinated fashion. They themselves were almost free of injuries. Nikkie did have a few superficial scratches and bites, but Yeroen had none at all, suggesting that he had held Luit down while letting the younger male inflict all the harm.

We will never know exactly what transpired, and unfortunately no females had been present to stop the fight. It is not unusual for them to collectively interrupt out-of-control male altercations. On the night of the assault, however, the females had been in separate night cages within the same building. They must have followed the entire commotion, but were helpless to intervene.

The colony was eerily silent that morning while Luit sat in his blood. It was the first time in the zoo's history that none of the apes ate their breakfast. After Luit was carried off and the rest of the colony let outside—onto a two-acre island with grass and trees—the first thing that happened was an unusually fierce attack on Nikkie by a female named Puist. She was so persistently aggressive that the young male, normally very impressive, fled up into a tree. On her own, Puist kept him there for at least ten minutes by screaming and charging every time he tried to come down. Puist had always been Luit's main ally among the females. Her night quarters offered a view into the males', and it seemed clear she was expressing her opinion about the deadly assault.

And so our chimpanzees had demonstrated all the elements of two-against-one politics, from the need for unity to the fate of a ruler who becomes too cocky. Power is the prime mover of the male chimpanzee. It's a constant obsession, offering great benefits if obtained and intense bitterness if lost.

## MALES ON PEDESTALS

Political murder is no more rare in our own species: John F. Kennedy, Martin Luther King, Salvador Allende, Yitzhak Rabin, Gandhi. The list goes on. Even the Netherlands—normally politically sedate (or "civilized," as the Dutch would say)—was shocked a few years ago by the assassination of Pim Fortuyn, a political candidate. Further back in history, my country saw one of the most gruesome political murders of all. Whipped into a frenzy by the adversaries of Johan de Witt, a mass of people got its hands on the statesman and his brother, Cornelius. They finished both men off with swords and muskets, hung their bodies upside down, and cut them open like pigs in a slaughterhouse. Hearts and innards were removed, grilled over a fire, and eaten by a festive mob! This ghastly event, which took place in 1672, resulted from profound frustration at a time when the country had lost a string of wars. The murder was commemorated in poems and paintings, and the Historical Museum of The Hague still exhibits a toe and the ripped out tongue of one of the victims.

For man or beast, death is the ultimate price of trying to reach the top. There is a chimp named Goblin from Gombe National Park in Tanzania. After many years as the group's bully, Goblin was set upon by a mass of angry chimpanzees. First, he lost a fight against a challenger who was supported by four younger males. As so often in the field, the actual fight was barely visible since it occurred in dense undergrowth. But Goblin emerged screaming, and fled with injuries on his wrist, feet, hands, and, worst of all, his scrotum. His wounds were strikingly similar to Luit's. Goblin nearly died as his scrotum became infected and started to swell, and he developed a fever. Days afterward, he was moving slowly, resting often, and eating little. But he was tranquilized with a dart by a veterinarian and given antibiotics. After a period of convalescence, during which he stayed out of sight of his own community, he tried to stage a comeback, directing intimidating charges at the new alpha male. This was a grave misjudgment, because it provoked a pursuit by the other males in the group. Seriously injured again, he was once more saved by the field veterinarian. Eventually, Goblin was accepted back into the community, but in a low-ranking position.

The fates that may befall those at the top are an inevitable part of the power drive. Apart from the risk of injury or death, being in a position of power is stressful. This can be demonstrated by measuring cortisol, a stress hormone in the blood. It is no easy task to do so in wild animals, but Robert Sapolsky has been darting male baboons on the African plains for years. Among these highly competitive primates, cortisol levels depend on how good an individual is at managing social tensions. As in humans, this turns out to be matter of personality. Some dominant males have high stress levels simply because they cannot tell the difference between a serious challenge by another male and neutral behavior that they shouldn't worry about. They are jumpy and paranoid. After all, if a rival walks by, it could be just because he needs to go from A to B, not because he wants to be a nuisance. When the hierarchy is in flux, misunderstandings accumulate, wrecking the nerves of males near the top. Since stress compromises the immune system, it's not unusual for high-ranking primates to develop the ulcers and heart attacks also common in corporate CEOs.

The advantages of high rank must be pretty enormous, otherwise evolution would never have installed such foolhardy ambitions. They are ubiquitous in the animal kingdom, from frogs and rats to chickens and elephants. High rank generally translates into food for females and mates for males. I say "generally," because males also compete for food, and females for mates, even though the latter is mostly restricted to species, like ours, in which males help out with child rearing. Everything in evolution boils down to reproductive success, which means that the different orientations of males and females make perfect sense. A male can increase his progeny by mating with many females while keeping rivals away. For the female, such a strategy makes no sense: mating with multiple males generally does not do her any good.

The female goes for quality rather than quantity. Most female animals do not live with their mates, hence all they need to do is pick the most vigorous and healthy sex partner. This way, their offspring will be blessed with good genes. But females of species in which the mates stay around are in a different situation, which makes them favor males who are gentle, protective, and good providers. Females further enhance reproduction by what they eat, especially if they are pregnant or lactating, when caloric intake increases fivefold. Since dominant females can claim the best food, they raise the healthiest offspring. In some species, like rhesus macaques, the hierarchy is so strict that a dominant female will simply stop a subordinate walking by with bulging cheek pouches. These pouches help the monkeys carry food to a safe spot. The dominant will hold the head of the subordinate and open her mouth, essentially picking her pocket. Her intrusion meets with no resistance because for the subordinate it's either this or get bitten.

Do the benefits of being on top explain the dominance drive? Looking at the outsized canine teeth of a male baboon or the bulk and muscle of a male gorilla, one sees fighting machines evolved to defeat rivals in pursuit of the one currency recognized by natural selection: offspring produced. For males, this is an all-or-nothing game; rank determines who will sow his seed far and wide and who will sow no seed at all. Consequently, males are built to fight, with a tendency to probe rivals for weak spots, and a certain blindness to danger. Risk-taking is a male characteristic, as is the hiding of vulnerabilities. In the male primate world, you don't want to look weak. So it's no wonder that in modern society men go to the doctor less often than women and have trouble revealing their emotions even with an entire support group egging them on. The popular wisdom is that men have been socialized into hiding emotions, but it seems more likely that these attitudes are the product of being surrounded by others ready to seize any opportunity to bring them down. Our ancestors must have noticed the slightest limp or loss of stamina in others. A high-ranking male would do well to camouflage impairments, a tendency that may have become ingrained. Among chimpanzees it's not unusual for an injured leader to double the energy he puts into his charging displays, thus creating the illusion of being in perfect shape.

Genetic traits that help males lay claim to fertile females will be passed on. Animals don't think in terms of procreation, but they do act in ways that help spread their genes. The human male has inherited the same tendency. There are plenty of reminders of the connection between power and sex. Sometimes, like during the Monica Lewinsky scandal, this connection is exposed with great fanfare and hypocrisy, but most people are realistic about the sex appeal of leaders and ignore their philandering. That said, this applies to male leaders only. Since men don't fancy powerful mates, high status fails to benefit women in the sexual domain. A prominent French politician once compared power to pastry—she loved it, but she knew it wasn't good for her.

These differences between the sexes emerge early on. A Canadian study invited boys and girls aged nine and ten to play games that measured competitiveness. Girls were reluctant to take toys away from each other unless it was the only way to win, but boys claimed toys regardless of how this affected the game's outcome. Girls competed only if necessary, but boys seemed to do so just for the sake of it.

Similarly, upon meeting for the first time, men check each other out by picking something—*anything*—to fight over, often getting worked up about a topic they normally don't care about. They adopt threatening body postures—legs apart and chests pushed out—make expansive gestures, speak with booming voices, utter veiled insults, make risqué jokes, and so on. They desperately want to find out where they stand relative to one another. They hope to impress the others sufficiently that the outcome will be in their favor.

This is a predictable event on the first day of an academic gathering when egos from the far corners of the globe face each other in a seminar room or, for that matter, at a bar. Unlike the women, who tend to stay on the sidelines, the men get so involved in the ensuing intellectual jostle that they sometimes turn red or white. What chimpanzees do with charging displays—with their hair on end, drumming on anything that amplifies sound, uprooting little trees as they go—the human male does in the more civilized manner of making mincemeat of someone else's arguments or, more primitively, giving others no time to open their mouths. Clarification of the hierarchy is a top priority. Invariably, the next encounter among the same men will be calmer, meaning that something has been settled, though it's hard to know what exactly that is.

For males, power is the ultimate aphrodisiac, and an addictive one at that. The violent reaction of Nikkie and Yeroen to their loss of power fits the frustration-aggression hypothesis to the letter: the deeper the bitterness, the greater the anger. Males jealously guard their power, and lose all inhibition if anyone challenges it. And this hadn't been the first time for Yeroen. The ferocity of the attack on Luit may have been due to the fact that it was the second time he had come out on top.

The first time Luit gained the upper hand—marking the end of Yeroen's ancient regime—I was perplexed by the way the established leader reacted. Normally a dignified character, Yeroen became unrecognizable. In the midst of a confrontation, he would drop out of a tree like a rotten apple, writhing on the ground, screaming pitifully, and waiting to be comforted by the rest of the group. He acted much like a juvenile ape being pushed away from his mother's teats. And like a juvenile who during tantrums keeps an eye on mom for signs of softening, Yeroen always noted who approached him. If the group around him was big and powerful enough, and especially if it included the alpha female, he would gain instant courage. With his supporters in tow, he would rekindle the confrontation with his rival. Clearly, Yeroen's tantrums were yet another example of deft manipulation. What fascinated me most, however, were the parallels with infantile attachment, nicely captured in expressions like "clinging to power" and "being weaned from power." Knocking a male off his pedestal gets the same reaction as yanking the security blanket away from a baby.

When Yeroen finally lost his top spot, he would often sit staring into the distance after a fight, an empty expression on his face. He was oblivious to the social activity around him and refused food for weeks. We thought he was sick, but the veterinarian found nothing wrong. Yeroen seemed a mere ghost of the impressive big shot he had been. I've never forgotten this image of a beaten and dejected Yeroen. When power was lost, the lights in him went out.

I've witnessed only one other such drastic transformation, this time in my own species. A senior professor, a colleague of mine on a university faculty, with extraordinary prestige and ego had failed to notice a budding conspiracy. Some young faculty members disagreed with him on a politically sensitive issue and successfully rallied a vote against him. Until then, I don't think anyone had ever had the guts to go head-to-head with him. Support for the alternative proposal had been cultivated behind his back by some of his own protégés. Following the fatal vote, which must have come out of the blue, given his expression of disbelief, all color drained from the professor's face. Looking ten years older, he had the same empty, ghostlike appearance Yeroen had after he had lost his top spot. For the professor, this was about much more than the issue at hand; it was about who ran the department. In the weeks and months following the meeting, his entire demeanor changed as he strode the corridors. Instead of saying "I am in charge," his body language now said "Leave me alone."

Bob Woodward and Carl Bernstein's *The Final Days* describes President Richard Nixon's breakdown after it became obvious that he would have to resign, "Between sobs, Nixon was plaintive. How had a simple burglary . . . done all this? . . . Nixon got down on his knees. . . . [He] leaned over and struck his fist on the carpet, crying, 'What have I done? What has happened?'" Henry Kissinger, his secretary of state, reportedly comforted the dethroned leader like a child. He consoled him, literally holding Nixon in his arms, reciting all of his great accomplishments over and over until the president finally calmed down.

## AN ARCHAIC TENDENCY

Given the obvious "will to power" (as Friedrich Nietzsche called it) of the human race, the enormous energy put into its expression, the early emergence of hierarchies among children, and the childlike devastation of grown men who tumble from the top, I'm puzzled by the taboo with which our society surrounds this issue. Most psychology textbooks do not even mention power and dominance, except in relation to abusive relationships. Everyone seems in denial. In one study on the power motive, corporate managers were asked about their relationship with power. They did acknowledge the existence of a lust for power, but never applied it to themselves. They rather enjoyed responsibility, prestige, and authority. The power grabbers were *other* men.

Political candidates are equally reluctant. They sell themselves as public servants, only in it to fix the economy or improve education. Have you ever heard a candidate admit he wants power? Obviously, the word "servant" is doublespeak: does anyone believe that it's only for our sake that they join the mudslinging of modern democracy? Do the candidates themselves believe this? What an unusual sacrifice that would be. It's refreshing to work with chimpanzees: they are the honest politicians we all long for. When political philosopher Thomas Hobbes postulated an insuppressible power drive, he was right on target for both humans and apes. Observing how blatantly chimpanzees jockey for position, one will look in vain for ulterior motives and expedient promises.

I was not prepared for this when, as a young student, I began to follow the dramas among the Arnhem chimpanzees from an observation window overlooking their island. In those days, students were supposed to be anti-establishment, and my shoulder-long hair proved it. We considered power evil and ambition ridiculous. Yet my observations of the apes forced me to open my mind to seeing power relations not as something bad but as something ingrained. Perhaps inequality was not to be dismissed as simply the product of capitalism. It seemed to go deeper than that. Nowadays, this may seem banal, but in the 1970s human behavior was seen as totally flexible: not natural but cultural. If we really wanted to, people believed, we could rid ourselves of archaic tendencies like sexual jealousy, gender roles, material ownership, and, yes, the desire to dominate.

Unaware of this revolutionary call, my chimpanzees demonstrated the same archaic tendencies, but without a trace of cognitive dissonance. They were jealous, sexist, and possessive, plain and simple. I didn't know then that I'd be working with them for the rest of my life or that I would never again have the luxury of sitting on a wooden stool and watching them for thousands of hours. It was the most revelatory time of my life. I became so engrossed that I began trying to imagine what made my apes decide on this or that action. I started dreaming of them at night and, most significant, I started seeing the people around me in a different light.

I am a born observer. My wife, who does not always tell me what she buys, has learned to live with the fact that I can walk into a room and within seconds pick out anything new or changed, regardless of how small. It could be just a new book inserted between other books or a new jar in the refrigerator. I do so without any conscious intent. Similarly, I like to pay attention to human behavior. When picking a seat in a restaurant I want to face as many tables as possible. I enjoy following the social dynamics—love, tension, boredom, antipathy—around me based on body language, which I consider more informative than the spoken word. Since keeping track of others is something I do automatically, becoming a fly on the wall of an ape colony came naturally to me.

My observations helped me see human behavior in an evolutionary light. By this, I mean not just the Darwinian light one hears so much about, but also the apelike way we scratch our heads if conflicted, or the dejected look we get if a friend pays too much attention to someone else. At the same time, I began to question what I'd been taught about animals: they just follow instinct; they have no inkling of the future; everything they do is selfish. I couldn't square this with what I was seeing. I lost the ability to generalize about "the chimpanzee" in the same way that no one ever speaks about "the human." The more I watched, the more my judgments began to resemble those we make about other people, such as this person is kind and friendly, and that one is self-centered. No two chimpanzees are the same.

It's impossible to follow what's going on in a chimp community without distinguishing between the actors and trying to understand their goals. Chimpanzee politics, like human politics, is a matter of individual strategies clashing to see who comes out ahead. The literature of biology proved of no help in understanding the social maneuvering, due to its aversion to the language of motives. Biologists don't talk about intentions and emotions. So I turned to Niccolò Machiavelli. During quiet moments of observation, I read from a book published four centuries earlier. *The Prince* put me in the right frame of mind to interpret what I was seeing on the island, though

I'm pretty sure that the philosopher himself never envisioned this particular application of his work.

Among chimpanzees, hierarchy permeates everything. When we bring two females inside the building—as we often do for testing—and have them work on the same task, one will be ready to go while the other hangs back. The second female barely dares to take rewards and won't touch the puzzle box, computer, or whatever else we're using in the experiment. She may be just as eager as the other, but defers to her "superior." There is no tension or hostility, and out in the group they may be the best of friends. One female simply dominates the other.

In the Arnhem colony, the alpha female, Mama, did occasionally underline her position with fierce attacks on other females, but she was generally respected without contest. Mama's best friend, Kuif, shared her power, but this was nothing like a male coalition. Females rise to the top because everyone recognizes them as leaders, which means there is little to fight over. Inasmuch as status is largely an issue of personality and age, Mama did not need Kuif. Kuif shared in, but did not contribute to, Mama's power.

Among the males, in contrast, power is always up for grabs. It's not conferred on the basis of age or any other trait, but has to be fought for and jealously defended in the face of contenders. If males form coalitions, it's because they need each other. Status is determined by who can beat whom, not just on an individual basis but in the group as a whole. It does not do a male any good if he can physically defeat his rival if each time he tries to do so the whole group jumps on top of him. In order to rule, a male needs both physical strength and buddies who will help him out when a fight gets too hot. When Nikkie was alpha, Yeroen's assistance was crucial. Not only did Nikkie need the old male's help to keep Luit in check, but he was also unpopular with the females. It was not unusual for females to band together against him. Yeroen, being highly respected, could stop such mass discontent by positioning himself between Nikkie and the screaming females. Nikkie's dependence makes it all the more surprising that he ended up biting the hand that fed him.

But with complex strategies come miscalculations. This is why we speak of political "skills": it's not so much who you are, but what you do. We are exquisitely attuned to power, responding quickly to any new configuration. If a businessman tries to get a contract with a large corporation, he will be in meeting after meeting with all sorts of people from which a picture emerges of rivalries, loyalties, and jealousies within the corporation he is visiting, such as who wants whose position, who feels excluded by whom, and who is on his way down or out. This picture is at least as valuable as the organizational chart of the company. We simply could not survive without our sensitivity to power dynamics.

Power is all around us, continuously confirmed and contested, and perceived with great accuracy. But social scientists, politicians, and even laypeople treat it like a hot potato. We prefer to cover up underlying motives. Anyone who, like Machiavelli, breaks the spell by calling it like it is, risks his reputation. No one wants to be called "Machiavellian," even though most of us are.

## GROVELING IN THE DUST

It's hard to name a single discovery in animal behavior that enjoys wider name recognition than the "pecking order." Even if pecking is not exactly a human behavior, the term is ubiquitous in modern society. We speak of the corporate pecking order, or the pecking order at the Vatican (with "primates" on top), acknowledging inequalities and their ancient origins. We're also mocking ourselves, hinting that we—sophisticated human beings—share a few things with domestic fowl.

It is something a child can see, and I mean this quite literally. The momentous discovery of the pecking order was made at the beginning of the twentieth century by a Norwegian boy, Thorleif Schjelderup-Ebbe, who fell in love with chickens at the tender age of six. His mother bought him his own flock, and soon each bird had a name. By the age of ten, Thorleif was keeping detailed notebooks, which he maintained for years. Apart from keeping track of how many eggs the chickens laid, and who pecked whom, he was particularly fascinated by the occasional exceptions to the hierarchy—"triangles" in which hen A is master over B, and B over C, but C over A. So from the start, like a real scientist, the child was not only interested in the regularities but also the irregularities of the rank order. We now find the social ladder that young Thorleif discovered—and later wrote up for his dissertation—so obvious that we cannot even imagine that anyone could miss it.

Likewise, watching a group of people, one will quickly notice which individuals act with the greatest confidence, attract the most glances and nods of agreement, are least reluctant to break into the discussion, speak in a softer voice yet expect everybody to listen (and laugh at their jokes!), voice unilateral opinions, and so on. But there are far more subtle status clues. Scientists used to consider the frequency band of 500 hertz and below in the human voice as meaningless noise, because when a voice is filtered, removing all higher frequencies, one hears nothing but a low-pitched hum. All words are lost. But then it was found that this low hum is an unconscious social instrument. It is different for each person, but in the course of a conversation people tend to converge. They settle on a single hum, and it is always the lower status person who does the adjusting. This was first demonstrated in an analysis of the *Larry King Live* television show. The host, Larry King, would adjust his timbre to that of high-ranking guests, like Mike Wallace or Elizabeth Taylor. Low-ranking guests, on the other hand, would adjust their timbre to that of King. The clearest adjustment to King's voice, indicating lack of confidence, came from former Vice President Dan Quayle.

The same spectral analysis has been applied to televised debates between U.S. presidential candidates. In all eight elections between 1960 and 2000 the popular vote matched the voice analysis: the majority of people voted for the candidate who held his own timbre rather than the one who adjusted. In some cases, the differences were extreme, such as between Ronald Reagan and Walter Mondale. And only in 2000 did a candidate with a slightly subordinate voice pattern, George W. Bush, get elected. But this was not really an exception to the rule because, as Democrats will relish pointing out, the popular vote actually went to the candidate with the dominant voice pattern, Al Gore.

Below the radar of consciousness, we thus communicate status every time we talk with someone, whether in person or on the telephone. On top of this, we have ways of making the human hierarchy explicit, from the size of our offices to the price of the clothes we wear. In an African village, the chief has the largest hut and a golden robe, and at university commencement ceremonies, professors in academic regalia proudly march past students and their parents. In Japan, the depth of the greeting bow signals precise rank differences not only between men and women (with women bowing more deeply), but also between senior and junior family members. Hierarchy is most institutionalized in male bastions such as the military, with its stars and stripes, and the Roman Catholic Church, where the pope dresses in white, cardinals in red, monsignors in purple, and priests in black.

Chimpanzees are every bit as formal as the Japanese in their greeting ceremonies. The alpha male makes an impressive display, going around with his hair on end, hitting anyone who doesn't move out of the way in time. The display both draws attention to the male and impresses his audience. One alpha male in the Mahale Mountains National Park in Tanzania developed the habit of dislodging enormous boulders so he could roll them down a dry riverbed, producing a thunderous noise. One can imagine the awe with which others watched a spectacle that they couldn't match. The performer would then sit down, waiting for his audience to approach. And they did, at first reluctantly, but then in droves, bowing—known as "bobbing"—and groveling, noisily proclaiming their respect with panting grunts. Dominant males seem to keep track of these greetings, because during their next round of display they sometimes single out parties who failed to acknowledge them for "special treatment" to make sure that next time they won't forget to greet.

I once visited the Forbidden City in Beijing—four times the size of Versailles, ten times that of Buckingham Palace—with its beautifully decorated buildings surrounded by gardens and vast squares. It wasn't hard to imagine Chinese emperors ruling from elaborate thrones designed to overlook enormous squatting masses, intimidating them with their splendor. European royalty still drive through the streets of London and Amsterdam in gold-plated carriages in a display of power that, although now largely symbolic, underlines the social order. The Egyptian pharaohs impressed audiences in a glorious ceremony that could be held only on the longest day of the year. They would stand in a specific spot in the Solar Temple of Amen-Ra, so that the sun would shine down the narrow hallway behind them, bathing them in such brilliance that it blinded their audience, confirming the rulers' divinity. On a more modest scale, prelates in colorful robes hold out their hands to subordinates who kiss their rings, and the queen is greeted by women with a special curtsy. But the prize for the weirdest status ritual goes to Saddam Hussein, the deposed tyrant of Iraq, who had his underlings greet him with a kiss on his armpit. Was the idea to give them a whiff of power?

Humans remain sensitive to physical markers of status. Diminutive men such as American presidential candidate Michael Dukakis or Italian prime minister Silvio Berlusconi ask for a box to stand on during debates and official group photos. There are pictures of Berlusconi smiling face-to-face with a leader whom he normally would see only at chest level. We may joke about their Napoleon complexes, but short men do have to work harder for authority. The same physical prejudices that apes and children use to sort out their relations remain at play in the human adult world.

Few people are aware of nonverbal communication, but one innovative business course draws attention to it by using dogs as "mirrors" for managers.

The managers give orders to the dogs, whose reactions tell them how convincing they are. The perfectionist, who tries to plan every step and is upset if something goes wrong, quickly loses a dog's interest, and people who give orders while their body language signals uncertainty will end up with a confused or skeptical dog. Not surprisingly, the optimal combination is one of warmth and firmness. Anyone who works with animals is used to their uncanny sensitivity to body language. My chimpanzees sometimes know my mood better than I do: it's hard to fool an ape. One reason for that is the absence of distraction by the spoken word. We attach such importance to verbal communication that we lose track of what our bodies say about us.

Neurologist Oliver Sacks described a group of patients in an aphasia ward convulsed with laughter during a televised speech by President Reagan. Incapable of understanding words as such, aphasia patients follow much of what is being said through facial expressions and body movement. They are so attentive to nonverbal clues that they cannot be lied to. Sacks concluded that the president, whose speech seemed perfectly normal to the nonpatients around, so cunningly combined deceptive words and tone of voice that only the brain damaged were able to see through it.

Not only are we sensitive to hierarchies and the body language associated with them, we simply could not live without them. Some people may wish them away, but harmony requires stability, and stability depends ultimately on a well-acknowledged social order. We can easily see what happens if stability is absent in a chimpanzee colony. Trouble starts when one male who used to go out of his way to bow and pay respect to the boss transforms himself into a defiant producer of noise and mayhem. He appears to grow in size, and every day makes charging displays a bit closer to the leader, demanding attention by hurling branches and heavy rocks in his direction.

At first, the outcome of such confrontations is open-ended. Depending on how much support each rival receives from others, a pattern will emerge, sealing the leader's fate if it turns out he has less support than his challenger. The critical moment is not the first victory for the challenger, but the first time he elicits submission. The former alpha may lose numerous run-ins, flee in panic, and end up screaming in the top of a tree, yet so long as he refuses to raise the white flag in the form of a series of low panting grunts accompanied by bowing to his challenger nothing has been settled.

The challenger, for his part, will not lighten up until the former alpha submits. In effect, the contender is telling the fallen king that the only way to be friends again will be to give the pantgrunts that admit defeat. It is pure blackmail; the challenger is waiting for the alpha to cry uncle. On many an occasion, I have seen a male who failed to pantgrunt when approaching a new alpha find himself standing alone. The alpha simply moves away—why bother with someone who doesn't recognize your position? It would be like a soldier greeting a superior officer without a salute. Appropriate respect is the key to relaxed relations. Only when rank issues have been settled will rivals reconcile and calm be restored.

The clearer the hierarchy, the less need for reinforcement. In chimpanzees, a stable hierarchy eliminates tensions so that confrontations become rare: subordinates avoid conflict and higher-ups have no reason to seek it. Everybody is better off. The group can hang out together, groom each other, play, and relax, because no one feels insecure. If I see chimpanzee males cavorting around with so-called play faces (a wide-open mouth accompanied by laughlike vocalizations), literally pulling each others' legs, nudging each other in jest, I know they are pretty sure who dominates whom. Since everything has been worked out, they can loosen up. As soon as one of them decides to challenge the existing order, though, play will be the first behavior to drop out. All of a sudden, they have more serious business to attend to.

Status rituals among chimpanzees are not just about power, therefore, they are also about harmony. The alpha male will imperiously stand with hair bristling after a perfect display, barely paying attention to subordinates prostrating themselves with respectful vocalizations, kissing his face, chest, or arms. By lowering his or her body and looking up at alpha, the pantgrunter makes it obvious who is on top, which paves the way for peaceful, friendly relations. And not only that, clarification of hierarchy is essential for effective collaboration. This is why the most cooperative human enterprises, such as large corporations and the military, have the best-defined hierarchies. A chain of command beats democracy any time decisive action is needed. We spontaneously switch to a more hierarchical mode depending on the circumstances. In one study, ten-year-old boys at a summer camp were divided into two groups that competed against each other. Out-group derogation—things like holding noses in disgust upon meeting members of the other group—quickly became common practice. On the other hand, in-group cohesiveness increased along with reinforcement of social norms and leader-follower behavior. The experiment demonstrated the binding quality of status hierarchies, which were reinforced as soon as concerted action was called for.

This brings me to the greatest paradox, which is that although positions within a hierarchy are born from contest, the hierarchical structure itself, once established, eliminates the need for further conflict. Obviously, those lower on the scale would have preferred to be higher, but they settle for the next best thing, which is to be left in peace. The frequent exchange of status signals reassures bosses that there is no need for them to underline their position by force. Even those who believe that humans are more egalitarian than chimpanzees will have to admit that our societies could not possibly function without an acknowledged order. We crave hierarchical transparency. Imagine the misunderstandings we would run into if people never gave us the slightest clue about their position in relation to us, either in terms of appearance or in how they introduce themselves. Parents would walk into their child's school and might just as easily be talking with the janitor as the principal. We would be forced to continuously probe others while hoping not to offend the wrong person.

It would be like inviting clerics to a gathering in which an all-important decision needs to be made while asking them to dress in identical garb. With ranks ranging from priest to pope, nobody would be able to tell who is who. The result would probably be an indecorous melee with the higher "primates" forced into spectacular intimidation displays—swinging from the chandeliers perhaps—to make up for the lack of color coding.

## GIRL POWER

Every schoolboy learns that members of the other "species"—the one he never plays with—can be teased and provoked only if there are not too many of them around. They tend to defend themselves as a group.

Female unity in the face of adversity is an ancient trait. I have already described how female gorillas straighten out a new male by resisting his charges and going after him together. Female chimpanzees, too, attack males as a group, especially abusive ones. These coalitions can deliver such a beating that any male is in an understandable hurry to get out of their way. Since none of the females can match a male in speed and strength, solidarity is crucial. In the Arnhem colony, this solidarity added to Mama's authority, since she was the orchestrator in chief. Not only did all females recognize her as their leader but she was not averse to reminding them. If during male power struggles a female supported a contestant other than Mama's choice, there could be grave repercussions. The renegade would have something to think about while licking her wounds.

Female power is less obvious among chimps in the wild. Females tend to travel alone with their dependent offspring, forced apart by their search for food—the fruits and leaves on which they rely. Resources are too dispersed for an entire group to forage together. Their spread-out existence precludes females from forming the same kind of alliances they have in confinement where the screams of one female recruit all others. Proximity reduces the gender gap. Take the way female chimpanzees at zoos "confiscate" male weapons, which is unheard of in the field. A female will approach a male who is gearing up for a confrontation, sitting with his hair erect, swaying his body from side to side and hooting. Males may warm up like this for ten minutes before launching a charge. This offers females a window of opportunity to pry open his hands to remove his weapons—things like heavy sticks or rocks. And females have excellent reasons for doing so: males often take their frustrations out on them.

The relative gender equality at the zoo may be artificial, but it's highly instructive. It hints at a potential for female solidarity that few would have predicted from watching chimpanzees in the wild. And it is precisely this potential that has been realized by the chimpanzee's sister species. Bonobo females operate as a team in the forest, where they live in a richer environment that does permit communal travel. Bonobos gather in larger groups than chimpanzees, and as a result females are far more sociable. A long history of female bonding, expressed in lots of grooming and sex, has done more than erode male supremacy; it has actually turned the tables. The result is a fundamentally different order. And yet at the same time I see continuity: female bonobos have perfected the female solidarity latent in all the African great apes.

The collective rule of female bonobos is well-known at zoos, and fieldworkers must have begun to suspect the same years before. But no one wanted to be the first to make such an outrageous claim, given how much male dominance is taken for granted in human evolution. Until 1992, that is, when scientists presented findings that left little doubt about bonobo girl power. One report looked at food competition in zoos, documenting how a male chimp living with two females will claim everything for himself, whereas a bonobo male under the same circumstances may not even be able to get near the food. He can make as many charging displays as he wants, but the females ignore the commotion and divide the food among themselves.

In the wild, an alpha female bonobo will stride into a clearing dragging a branch behind her, making a display that is avoided and watched by all others. It's not unusual for female bonobos to chase off the males, laying claim to the large fruits they divide among themselves. *Anonidium* fruits weigh up to twenty-two pounds, and *treculia* fruits up to sixty-six pounds, nearly the weight of a grown bonobo. After they have dropped to the ground, these colossal fruits are claimed by females, who only sometimes see fit to share with begging males. Whereas it's not rare for individual male bonobos to supplant individual females, especially younger ones, females will always collectively dominate the males.

Given our own fascination with gender issues, it's no wonder that bonobos became an instant hit. Alice Walker dedicated *By the Light of My Father's Smile* to our close kinship, and *New York Times* columnist Maureen Dowd once mixed political commentary with praise for the bonobo's sexual equality. To others, however, the bonobo seems almost too good to be true. Could it be a politically correct concoction, a made-up ape meant to satisfy liberals? Some scientists insist that male bonobos are not subordinate but simply "chivalrous." They speak of their "strategic deference," thus ascribing the weaker sex's clout to the stronger sex's good-heartedness. After all, they note, female dominance seems limited to food. Others try to push bonobos out of the human ancestry picture altogether. One well-known anthropologist went so far as to suggest that bonobos can be safely ignored since they're endangered in the wild, the implication being that only successful species are worth considering.

Are male bonobos just nice guys? If there is one yardstick that has been applied to every animal on the planet, it is that if individual A can chase individual B away from its food, A must be dominant. As noted by Takayoshi Kano, the Japanese scientist who has studied bonobos for twenty-five years in Africa, food is exactly what female dominance is all about. If this is what matters to them, it should matter to the human observer as well. Kano went on to point out that even if there is no food around, fully grown males react with submission and fear to the mere approach of a top female.

Among those who work with bonobos, the initial shock and disbelief have worn off. We have gotten so used to the upside-down order of the sexes that we cannot even imagine that it could be otherwise. It seems entirely natural. The skeptics apparently can't get beyond the way things work in our species. During the tour for my book *Bonobo: The Forgotten Ape*, the high point—or perhaps it was the low point—was a question posed by a highly respected German biology professor. He stood up after my lecture and barked in an almost accusatory tone, "What's wrong with those males?!" He was shocked by the female dominance. I, on the other hand, have always thought that, given the bonobo's high rate of sexual activity and low aggression levels, the males don't have much to complain about. One would assume that they are less stressed than their human and chimpanzee brothers. My response to the professor—that bonobo males seemed to be doing just fine—did not appear to satisfy him, however. This ape profoundly challenges assumptions about our lineage and behavior.

So what's so good about being a male bonobo? For one thing, the male/female ratio among adults in the wild is almost one to one. Bonobo societies include equal numbers of males and females, whereas chimp societies often include twice as many females as males. Since both species have a one to one sex ratio at birth, and since there are no roaming males outside the community, chimp males must suffer extraordinary mortality. This is hardly surprising, given the intercommunity warfare of this species as well as the injuries and stress associated with continuous power struggles. The upshot is that male bonobos lead longer, healthier lives than their macho counterparts.

There was a time when bonobos were thought to have a family structure like our own: adult males were found to have stable bonds with particular females. Finally, an ape to enlighten us about the origins of monogamy, we thought. But then we learned from the patient fieldwork of Kano and others that those bonds were actually between mothers and sons. A fully grown male follows his mother around through the forest, benefiting from her attention and protection, especially if she is of high rank. In fact, the male hierarchy is a maternal affair. Instead of forming ever-changing coalitions among themselves, male bonobos vie for positions on their mothers' apron strings.

A typical example is a wild alpha female, Kame, with no fewer than three grown sons, the oldest of whom was alpha male. When old age began to weaken Kame, she became hesitant to defend her children. The son of the beta female must have sensed this because he began to challenge Kame's sons. His own mother backed him up and was not afraid to attack the top male on his behalf. The frictions escalated to the point of the two mothers exchanging blows, rolling on the ground, with the beta female holding Kame down. Kame never recovered from this humiliation, and soon her sons dropped to mid-ranking positions. After Kame's death they became peripheral, and the son of the new alpha female took the top spot.

Had Kame's sons been chimpanzees, they would have rallied together in defense of their positions. In bonobos, however, male alliances are poorly developed, which is precisely what permits females to have such an impact. Even if rare, observations of such power struggles also give the lie to the idea that bonobo society is purely egalitarian. Tensions are by no means absent, males are highly competitive, and females can be as well. Rank seems to pay off in a big way. Top males enjoy greater female tolerance around food and have more sex partners. This means that a mother who manages to get a son into the upper echelons promotes her own lineage via the grandchildren he will sire. Not that bonobos understand this connection, but natural selection must have favored mothers who actively assist their son's status quests.

Does this mean that bonobo society is basically chimp society in reverse? Hardly. To my mind, the chimpanzee is far more of a *zoon politikon* (political animal). This has to do with the way coalitions are formed as well with as the different nature of the female hierarchy. In both apes and humans, the female hierarchy is less contested and consequently requires less enforcement. Women rarely think about themselves in hierarchical terms, and their relations are never as formalized as those among men. But undeniably there are women who command more respect than others. It is far more common for older women to dominate younger ones than the reverse. Within a single social stratum, senior women seem to rule. Traditionally, women exert their greatest influence in the family, where they don't need to fight, boast, or bluff their way to the top: they simply get there with age. Personality, education, and family size certainly matter, and there are many subtle ways in which women compete, but all else being equal, seniority seems at least half the story when it comes to a woman's position among other women.

The same applies to apes. In the field, older females have a leg up over younger ones who arrive freshly from the outside. Females leave their community at puberty to join another. Chimpanzee females need to find a range for themselves in the territory of their new community, often in competition with resident females. Young bonobo females, with their closer female ties, seek "sponsorship" from a resident, grooming her and having sex with her, after which the older female acts as protector of the younger one, taking her under her wing. With time, the young female herself may become a sponsor of new immigrants, thus repeating the cycle. This system, too, is biased toward seniority. Even if female hierarchies are never perfectly age-graded, age goes a long way toward explaining the order.

Dominance struggles among female apes are far less common than among males. And if they do occur, they are always among females from the same age bracket. In a group that includes females over thirty years old, one will never find a twenty-year-old at the top. This is not a matter of physical prowess—twenty-year-olds are at their peak—but there seems a total absence of willpower in younger females to take on one of those experienced, hard-as-nails older ladies. I know alpha females whose positions have remained uncontested for decades. Obviously, there is a limit to a top female's staying power, depending on both her physical and mental health, but females reach this point decades later than males.

The way older females put younger ones in their place is fascinating, since most of the time it's done without any overt aggression. Viewed as mother figures by the younger ones, who don't have their own mothers around, all that an older female needs to do to send a message is reject an overture, refuse to share food, or turn around and walk away from a grooming attempt. The older female is putting the emotional screws on. The younger one may throw a tantrum, which the older one will watch without any perturbation: she has seen this before. The reasons for snubs are often subtle as well. A snub may occur hours after the younger female pinched an offspring of the older female, grabbed a piece of food that the older female had been heading for, or failed to leave the alpha male when the older female came over to groom him. For the human observer, however, female interactions are obviously harder to track than the straightforward confrontations among males.

With male dominance based on fighting abilities and support from friends, the impact of age on male hierarchies is quite different. Getting older is never to a male's advantage. Alpha males rarely stay in power for more than four or five years. In a male-ruled system, like the chimp's, top positions become vacant on a regular basis, whereas in a female-ruled system, like the bonobo's, social change is both less common and more gradual. Only if the alpha female weakens or dies will there be movement, and then only near the top. This creates considerably less room for ambitious individuals to improve their positions.

There is another reason why there is less political maneuvering among bonobos: their coalitions depend on kinship. Like age, kinship is a given: sons cannot pick their mothers. The male bonobo needs to be alert to opportunities to climb the social ladder and in this sense is no less competitive than the male chimpanzee. But since everything depends on his mother and her position vis-à-vis other females, the male bonobo also needs to be patient. He has less opportunity to shape his own future than the male chimp, who is free to enter into a range of partnerships with other males. Some of these males may be his brothers, but he can also align himself with unrelated males. Thanks to this far more flexible situation, male chimps have evolved into opportunistic strategists, endowed by nature with an appropriately aggressive temper and intimidating physique. They pack enormous muscles, looking coarse and menacing next to bonobo males with their lightweight bodies and more sensitive expressions.

Life in a matrifocal society has, therefore, created a different kind of male. There is nothing wrong with the bonobo male, even though most men would not want to be like him. He lacks the sort of control over his own destiny that males of his nearest relatives, humans and chimpanzees, claim as their birthright.

## STRENGTH IS WEAKNESS

When tensions rise, male chimpanzees stay close together. This is why Yeroen, Luit, and Nikkie shared a cage on that fatal night. The caretakers and I wanted each male to sleep alone, but it's hard to control animals as strong as chimps. As soon as two of them entered a night cage together, the third absolutely wanted to join them. He simply could not afford to be left out. How could Luit prevent a hostile axis if he was not around to keep the other two from grooming each other? The evening before his death, we spent hours trying to separate the three males, but to no avail. It was as if they were glued together, sneaking as a team through open doors, holding on to each other's hips, so that no one was ever left behind. We resigned ourselves to leaving them together for the night.

Two-against-one dynamics are a familiar problem in human families with triplets, where one triplet is often left out of games by the other two. The lore among hunting peoples says that men should never set off in a party of three since only two of them may return alive (meaning that two will turn against the third). We easily grasp triadic configurations. In chess, a rook and a bishop can hold their own against a queen, and in real life we ask a friend to put in a good word for us so that we do not stand alone.

Male chimps are intimately familiar with this dynamic and seem to realize the importance of their own coalitions. Infighting among coalition partners is so threatening that they desperately try to reconcile, especially the one who stands to lose the most, which often is the one at the top. Yeroen and Nikkie were always in a hurry to make up after a fight: they needed to preserve a united front. One moment they'd be running around screaming at each other, usually in competition over a female, and the next they'd fling their arms around each other and make up with a kiss. This signaled to everyone else that they intended to stay in power. The day they failed to reconcile was the day both of them dropped in rank.

The same phenomenon occurs between battling candidates in a political party. After one has emerged as the party's candidate, the loser rushes to endorse him. No one wants the opposition to think that the party is in tatters. They pat each other on the back, two former enemies now smiling at the cameras together. After George W. Bush had won the bitterly contested 2000 Republican nomination, his rival, John McCain, with a forced smile, faced reporters who expressed doubt that he was ready to forgive and forget. McCain got a big laugh saying over and over, "I endorse Governor Bush, I endorse Governor Bush, I endorse Governor Bush."

Coalition politics take place at the international level as well. I was once invited to a think tank in Washington, D.C. Our group was an interesting mix of policy makers, anthropologists, psychologists, Pentagon types, political scientists, and one primatologist (me!). It was shortly after the fall of the Berlin Wall. This historic event had meant a lot to me. When I lived in the Netherlands, the Soviet occupiers of East Germany could have been at my doorstep in two hours, something I was reminded of every time heavy NATO military vehicles rumbled down a nearby highway.

The underlying assumption at the meeting was that we were going to be living in a safer world now that one of the world's two largest military powers was fading into the past. Our task was to discuss what to expect: how would the new world order work out and what kinds of good things could the United States do with its newfound status as the sole superpower. I had problems with the underlying premise, however, since the demise of one power does not necessarily give the other free rein. This might be true in a simpler world, but Americans sometimes forget that their country is home to less than 4 percent of the planet's population. My evaluation of the international situation would have been easy to ignore, based as it was on animal behavior, had one of the political scientists not made the same points, but based on military history. Our message could be summarized in three deceptively simple words from coalition theory: strength is weakness.

This theory is nicely illustrated by Yeroen's choice of partner after he lost his position. For a brief while, Luit was alpha. Since Luit was physically the strongest male, he could handle most situations by himself. Furthermore, soon after his rise, the females one by one switched over to his side, most important, Mama. Mama was pregnant at the time, and it's natural that females under such circumstances do everything to stabilize the hierarchy. Despite his cushy position, Luit was keen on disrupting get-togethers among other males, especially between Yeroen and the only male who could pose a threat, Nikkie. Sometimes these scenes escalated into fighting. Noticing that both other males wanted to be his buddy, Yeroen grew in importance by the day.

At this point, Yeroen had two choices: He could attach himself to the most powerful player, Luit, and derive a few benefits in return—what kind of benefits would be up to Luit. Or, he could help Nikkie challenge Luit and in effect create a new alpha male who would owe his position to him. We have seen that Yeroen took the second route. This is consistent with the "strength is weakness" paradox, which says that the most powerful player is often the least attractive political ally. Luit was too strong for his own good. Joining him, Yeroen would add little. As the colony's superpower, Luit really did not need more than the old male's neutrality. Throwing his weight behind Nikkie was a logical choice for Yeroen. He would be the puppet master, having far more leverage than he could ever have dreamt of having under Luit. His choice also translated into increased prestige and access to females. So if Luit demonstrated the "strength is weakness" principle, Yeroen illustrated the corresponding "weakness is strength" principle according to which minor players can position themselves at an intersection that offers great advantage.

The same paradoxes operate in the international political arena. Ever since Thucydides wrote about the Peloponnesian War more than two millennia ago, it has been known that nations seek allies against nations perceived as a common threat. Fear and resentment drive weaker parties into each other's arms, making them weigh in on the lighter side of the balance. The result is a power balance in which all nations hold influential positions. Sometimes a single country is the main "balancer," as Great Britain was in Europe before World War I. Having a strong navy and being virtually immune to invasion, Britain was in the perfect position to prevent any continental power from gaining the upper hand.

Counterintuitive outcomes are not unusual. Think about a parliamentary system in which a majority vote out of one hundred is needed, and in which there are three parties, two with forty-nine seats each and one tiny party with only two seats. Which party do you think is the most powerful? Under such circumstances (which, in fact, existed in Germany during the 1980s), the party with two votes will be in the driver's seat. Coalitions are rarely larger than they need to be to win, hence the largest two parties have no desire to govern together. Both will court the smaller party, giving it disproportionate power.

Coalition theory also considers "minimally winning coalitions," in which players prefer to be part of a coalition that is large enough to be victorious yet small enough for them to make a difference within it. Inasmuch as siding with the strongest party dilutes payoffs, it's rarely the first choice. Even if in the foreseeable future the United States will be the most powerful player on the global stage, both economically and militarily, this by no means guarantees its inclusion in winning coalitions. On the contrary, resentment will build automatically, leading to counterbalancing coalitions among the remaining powers. It was coalition theory that I talked about at the think tank meeting, believing it was a widely accepted idea, but my comments were met with distinctly unhappy faces. The Pentagon clearly was not planning according to any "strength is weakness" scenario.

It did not take long, though, for precisely such a scenario to play out. One morning in the spring of 2003, I woke up to the unexpected sight in my newspaper of three smiling foreign ministers walking side by side toward the United Nations Security Council chambers. The ministers of France, Russia, and Germany had proclaimed their opposition to the planned United States led invasion of Iraq, noting that China, too, was on their side. There is no abundance of love between the French and Germans nor between the Chinese and Russians, but these strange bedfellows had come together after the U.S. government had abandoned consensus-building, which until then had permitted it to act as the world's most powerful player without upsetting international alliances. Isolation was setting in. The end of U.S. diplomacy had called forth a counter alignment that ten years earlier would have been unimaginable.

## MAGNA CARTA OF THE APES

Oddly enough, life below sea level explains the egalitarian ethos of the Dutch. The onslaught of storm floods in the fifteenth and sixteenth centuries instilled a sense of common purpose. The boy with his finger in the dike never existed. Every burgher had to pitch in to keep the country dry, carrying heavy clay bags in the middle of the night if a dike was near collapse. An entire city could be swallowed up in no time. Those who placed status above duty were frowned upon. Even today, the Dutch monarchy is ambivalent about pomp and circumstance. Once a year, the queen rides a bicycle and serves hot chocolate to her staff to show she is one of the people

The nature of hierarchies is culturally variable. It runs the gamut from the military formality of the Germans and the sharp class divisions of the

English to the Americans' easygoing attitudes and love of equality. But however relaxed some cultures are, nothing compares with the negation of status in what anthropologists call the true egalitarians. These people go well beyond having a queen who rides a bike or a president named Bill. The very idea of a monarch is offensive to them. I'm talking about the Navajo Indians, Hottentots, Mbuti Pygmies, !Kung San, Inuit, and so on. Ranging from hunter-gatherers to horticulturalists, these small-scale societies are said to completely eliminate distinctions—other than those between the sexes, or between parent and child—of wealth, power, and status. The emphasis is on equality and sharing. It is believed that our immediate ancestors lived like this for millions of years. Could it be, then, that hierarchies are less ingrained than we think?

There was a time when anthropologists saw egalitarianism as a passive, peace-loving arrangement in which people were at their best, loving and valuing each other. It was a utopian state where lion and lamb were said to sleep side by side. I'm not saying that such states are out of the question—in fact, a lioness on the Kenyan plains was recently reported to have been observed showering maternal affection on an antelope calf—but from a biological perspective, they're unsustainable. At some point, self-interest will rear its ugly head: predators will feel their empty stomachs and people will clash over resources. Egalitarianism is *not* based on mutual love and even less on passivity. It's an actively maintained condition that recognizes the universal human desire to control and dominate. Instead of denying the will to power, egalitarians know it all too well. They deal with it every day.

In egalitarian societies, men trying to dominate others are systematically undermined, and male pride is frowned upon. The proverbial fish tale is considered improper. Upon returning to his village, the successful hunter simply sits down in front of his hut without a word. He lets the blood on the shaft speak for itself. Any hint of boasting will be punished with jokes and insults about his miserable catch. Similarly, the would-be chief who gets it into his head that he can tell others what to do is openly told how amusing his airs are. Anthropologist Christopher Boehm studied these leveling mechanisms. He found that leaders who become bullies, are self-aggrandizing, fail to redistribute goods, and deal with outsiders to their own advantage quickly lose the respect and support of their community. If the usual tactics of ridicule, gossip, and disobedience fail, egalitarians are not beyond taking drastic measures. A Buraya chief who appropriated the livestock of other men and forced their wives into sexual relations was killed, as was a Kaupaku leader who overstepped his bounds. A good alternative, of course, is to simply desert a bad leader. All he will be left with is himself to boss around.

Since it's hard to survive without any leadership at all, egalitarians often permit certain men to act as first among equals. The keyword here is "permit," because the whole group will guard against abuses. In doing so, they employ social tools typical of our lineage, but ones we share with our primate relatives. Over the years, my team has recorded thousands of situations in which a third party intervenes in a fight to support one party or the other. We have done so with both monkeys and apes.

Monkeys tend to support winners, which means that dominant individuals rarely meet resistance. On the contrary, the group offers them a helping hand. No wonder monkeys have such strict and stable hierarchies. Chimps are fundamentally different in that they support losers as often as winners when intervening in a fight. An aggressor can never be sure, therefore, whether he will be aided or resisted. This is a crucial difference from a monkey society. The tendency of chimps to rally for the underdog creates an inherently unstable hierarchy in which the power at the top is shakier than in any monkey group.

A typical example occurred when Jimoh, the alpha male of our chimp group at the Yerkes Field Station, suspected a secret mating between an adolescent male and one of Jimoh's favorite females. The younger male and the female had wisely disappeared from view, but Jimoh had gone looking for them and had found the young male. Normally, the old male would merely chase off the culprit, but for some reason—perhaps because this female had refused to mate with him that day—he went full speed after him and did not relent. The pursuit went all around the enclosure, the young male screaming and literally having diarrhea from fear while Jimoh was intent on catching him.

Before the alpha could get to this point, however, females close to the scene began to "woaow" bark. With its indignant sound, this call is used in protest against aggressors and intruders. At first, the callers looked around to see how the rest of the group might react, but when others joined in, particularly the alpha female, the intensity of their calls increased until everyone's voice was heard in a deafening chorus. The scattered beginning gave the impression of the group taking a vote. But once the protest had swollen to a crescendo, Jimoh broke off his attack with a nervous grin on his face: he got the message. Had he failed to stop, there would undoubtedly have been concerted action to end the disturbance.

Punishment of bullying males may be severe. There have been reports of ostracism in wild chimpanzees, with males being forced to spend time in the danger zone at the territorial border—one report spoke of males "going into exile." Ostracism is usually precipitated by a mass attack, such as the one on Goblin at Gombe Stream, who was assaulted by a large coalition and might not have survived without veterinary treatment. Goblin was almost killed twice, leading field-workers to speculate that the violence with which he was ousted had to do with the nature of his rule, which they described as "tempestuous."

If those at the lower end of the social scale collectively draw a line in the sand, threatening serious consequences if those at the upper end step over it, we have the beginnings of what in legal terms is called a constitution. Obviously, present-day constitutions are full of refined concepts far too complex to apply to face-to-face human groups, let alone ape societies. Yet we should not forget that the U.S. Constitution, for example, was born from a revolution against the English sovereign. Its wonderful prose "We the people . . ." speaks with the voice of the masses. Its predecessor was the Magna Carta of 1215, in which King John's subjects threatened to wage war against and kill their oppressor if he did not drop his excessive appropriations. Again, the principle was one of collective resistance against an overbearing alpha male.

If top-ranking individuals can be so problematic, why have them at all? Well, to settle disputes, for one. Instead of having everybody take sides, what better way to handle the situation than by investing authority in a single person, a council of elders, or a government to serve the greater good by keeping order and finding solutions to disagreements? By definition, egalitarian societies lack the social hierarchy that could impose its will in disputes, hence they depend on arbitration. Impartiality is the key. Taken up by the judiciary in modern society, arbitration protects society against its greatest enemy: festering discord.

Dominant chimps generally break up fights by either supporting the weak against the strong or through impartial intervention. They may position themselves with all their hair on end between the combatants until they stop screaming; scatter them with a charging display; or literally pry locked fighters apart with both hands. In all of this, their main objective seems to be to put an end to the hostilities rather than to support one party or the other. For example, within weeks of attaining alpha rank Luit, the most even-handed leader I have known, adopted what is known as the "control role." A quarrel between two females got out of hand and ended with hair flying. Numerous apes rushed up to join in the fray. A huge knot of fighting, screaming apes rolled around in the sand until Luit leapt in and literally beat them apart. He did not choose sides, like everybody else had. Instead, *anyone* who continued to fight received a blow from him.

One might think that apes would support their relatives, friends, and allies. This is indeed true for most members of an ape society, but the male in control follows different rules. As alpha, Luit seemed to place himself *above* the conflicting parties, his intercessions aimed at the restoration of peace rather than at aiding his friends. Luit's interventions in conflicts on behalf of certain individuals did not correspond with how much time he spent sitting and grooming with them. He was the only impartial chimpanzee, meaning that he dissociated his job as arbiter from his social preferences. I have known other males to do the same, and when Christopher Boehm moved from anthropology to primatology, he also observed high-ranking chimps in the wild who were effective at heading off, stopping, or ameliorating conflicts.

A community does not accept the authority of every would-be arbiter. When Nikkie and Yeroen ruled the Arnhem colony as a team, Nikkie tried to step in when disputes occurred. More often than not, however, he ended up on the receiving end of the violence. Older females, especially, did not accept him coming in and hitting them over the head. One reason may have been that Nikkie was far from impartial: he sided with his friends regardless of who had started the fight. Yeroen's attempts at pacification, in contrast, were always accepted. In due time, the old male took over the control role from his junior partner. If a fight erupted, Nikkie would not even bother getting up, leaving its settlement to Yeroen.

This goes to show that the control role can be performed by the second-in-command and that the group has a say in who performs it. If the control role is an umbrella shielding the weak against the strong, it's held up by the community as a whole. Its members throw their weight behind the most effective arbitrator, providing him with the broad base needed to guarantee peace and order. This is important, because even the smallest squabble between two juveniles can escalate into something far worse. Juvenile fights induce tensions among mothers, with each mother inclined to protect her own offspring. In a pattern not unheard of in human day-care centers, the arrival of one mother on the scene raises the hackles of the second.

Having a higher authority take care of these problems—and being secure in the knowledge that he will do so with fairness and minimum force—is a relief for all.

What we see in the chimp, then, is a halfway station between the rigid hierarchies of monkeys, on the one hand, and the human tendency toward equality on the other. People never reach perfect equality, of course, not even in small-scale societies. And leveling the human hierarchy is a continuous struggle for the simple reason that we are born to strive for status. To the degree that it is achieved, egalitarianism requires subordinates to unite and watch over their own interests. The politicians themselves may be in it for the power, but the electorate is focused on their service. No wonder politicians would rather talk about the latter than the former.

When we elect leaders, we in effect tell them "You can be high up there in the capital so long as we find you useful." Democracy thus elegantly satisfies two human tendencies at once: the will to power and the desire to hold it in check.

## THE ELDER STATESMAN

I gave Mama her name because of her matriarchal position in the Arnhem chimpanzee colony. All the females obeyed her and all the males saw her as the final broker in political disputes. If tensions escalated to a point at which an actual fight became inevitable, arguing males would rush to her and sit in her arms, one on each side, screaming at each other. Great self-confidence combined with a maternal attitude put Mama at the absolute center of power.

She is still alive today. Every time I visit the zoo, she picks my face out from among the crowd of visitors and moves her arthritic bones closer to greet me from across the moat. In fact, she pantgrunts to me, meaning that she sees me as high-ranking even though I can assure you that in a fight with Mama, I would stand no chance. Undoubtedly, she realizes this, too. Far from letting this confuse us, however, we both know that the public structure is one thing and the reality of who can do what to whom quite another.

This double-layered nature of society is intriguing. Its formal structure must be transparent to serve its function, yet behind it we find murkier influences. An individual can be powerful without being at the top or, conversely, be at the top without having much sway. In Arnhem, for example, pantgrunting and bowing put Nikkie formally above Yeroen, Yeroen above Luit, Luit above Mama, Mama above all other females, and so on. Behind this clear order, however, existed a shadow framework, in which Yeroen held the strings controlling Nikkie, Luit's power was largely neutralized, and Mama wielded an influence that possibly surpassed Yeroen's.

We are adept at figuring out what's going on behind the scenes at our workplace and at realizing that following the social ladder to the letter will not get us anywhere. There are always high-ranking persons of little consequence and low-ranking ones (such as the boss's secretary) that one needs to befriend. The formal structure does get reinforced at times of crisis but, overall, we humans tend to establish a loose order of crisscrossing influences. We have expressions such as "the power behind the throne" and "figurehead" reflecting the same complexities found in a chimp colony.

In the Mahale Mountains, field-workers have seen older male chimps act as Yeroen did. As soon as a male is physically over the hill, he starts to play games, siding sometimes with one, sometimes with another of the younger males, thus making himself key to everybody's success. He turns his weakness into a strength. Elder statesmen in human politics come to mind: the post-prime, silver-haired Dick Cheneys and Ted Kennedys, who have given up any ambition to occupy the highest office themselves, but to whom others flock for counsel. Being too narrowly focused on their own careers, younger men are less useful advisers.

Jessica Flack spent hundreds of hours sitting on a tower in the hot Georgia sun focusing entirely on the pantgrunts with which chimpanzees acknowledge higher rank. She found that the top-ranking male does not necessarily get the most grunts. He receives them from his immediate rivals, which is what makes him the formal alpha, but the rest of the group may regularly pass alpha by to bow for, pantgrunt to, and kiss a different male. With alpha looking on, this is an awkward situation, but what's interesting is that these other males are invariably the ones who perform the role of arbiter in disputes. At Arnhem Zoo, too, we saw more pantgrunts aimed at Yeroen, the group's main arbiter, than at Nikkie, the actual boss. It's almost as if a group "votes" for popular mediators, bestowing respect upon them and thus upsetting alpha who, after having been repeatedly ignored, may start a spectacular bluff display to show that he counts, too.

By bringing Nikkie to power, Yeroen had carved out an influential role for himself. With Luit's death, however, his leverage evaporated. All of a sudden, Nikkie didn't need the old male anymore. Finally he could be boss on his own, or so he must have thought. Soon after I had left for America, however, Yeroen began to cultivate a tie with Dandy, a younger male. This took several years, but eventually led to Dandy challenging Nikkie as leader. The ensuing tensions drove Nikkie to a desperate escape attempt. He actually drowned trying to make it across the moat around the island. The local newspaper dubbed it a suicide, but to me it seemed more likely a panic attack with a fatal outcome. Since this was the second death on Yeroen's hands, I must admit that I've always had trouble looking at this scheming male without seeing a murderer.

A year after this tragic incident, my successor decided to show the chimps a movie. *The Family of Chimps* was a documentary filmed at the zoo when Nikkie was still alive. With the apes ensconced in their winter hall, the movie was projected onto a white wall. Would they recognize their deceased leader? As soon as a life-sized Nikkie appeared on the wall, Dandy ran screaming to Yeroen, literally jumping into the old male's lap! Yeroen had a nervous grin on his face. Nikkie's miraculous "resurrection" had temporarily restored their old pact.

## THE MONKEY'S BEHIND

Consciously or unconsciously, social dominance is always on our minds. We display typical primate facial expressions, such as retracting our lips to expose our teeth and gums when we need to clarify our social position. The human smile derives from an appeasement signal, which is why women generally smile more than men. In myriad ways our behavior, even at its friendliest, hints at the possibility of aggression. We bring flowers or a bottle of wine when invading other peoples' territories, and we greet each other by waving an open hand, a gesture thought to originate from showing the absence of weapons. We formalize our hierarchies—through body postures and tone of voice—to the point that an experienced observer can tell in only a few minutes who is high or low on the totem pole. We talk about human behaviors such as "ass-kissing," "groveling," and "chest-pounding" that constitute official behavioral categories in my field of study, suggesting a past in which hierarchies were acted out more physically.

Yet at the same time humans are inherently irreverent. It was Saint Bonaventura, a thirteenth-century theologian, who said, "The higher a monkey climbs, the more you see of its behind." We love to make fun of the higher-ups. We're always ready to bring them down. And the powerful know this all too well. "Uneasy lies the head that wears a crown," as Shakespeare wrote. China's first almighty emperor, Ch'in Shih Huang Ti, was so concerned about his safety that he covered all roads leading to his palaces so that he could come and go unnoticed. Nicolae Ceauşescu, the executed dictator of Romania, had three levels of labyrinthine tunnels, escape routes, and bunkers stocked with food constructed underneath the Communist Party building on the Boulevard of Socialist Victory in Bucharest.

Fear is obviously greatest in unpopular leaders. Machiavelli rightly noted that one is better off becoming Prince with the support of the common people than with the help of the nobility, because the latter feel themselves so close to your position that they will try to undermine it. And the broader your power base, the better. This is good advice for chimps, too: males who stand up for the oppressed are the most loved and respected. Support from below stabilizes the top.

Was democracy truly achieved via a hierarchical past? There is an influential school of thought which believes that we started out in a state of nature that was harsh and chaotic, ruled by the "law of the jungle." We escaped this by agreeing on rules and delegating enforcement of these rules to a higher authority. It is the usual justification of top-down government. But what if it was entirely the other way around? What if the higher authority came first and attempts at equality later? This is what primate evolution seems to suggest. There never was any chaos: we started out with a crystal clear hierarchical order and then found ways to level it. Our species has a subversive streak.

There are plenty of passively peaceful and tolerant animals. Some monkey species rarely bite each other, reconcile easily after fights, tolerate each other around food and water, and so on. The woolly spider monkey barely fights at all. Primatologists speak of different "dominance styles," meaning that the higher-ups are laid-back and tolerant in one species, but despotic and punitive in another. Even if some monkeys can be easygoing, however, they are never egalitarian. This would require that subordinates stage revolts and draw lines in the sand, which monkeys do only to a limited degree.

Bonobos are relaxed, too, and relatively peaceful. Employing the same leveling mechanisms as chimps, they have taken them to the extreme by actually turning the hierarchy upside down. Instead of stirring from below, the weaker sex acts from above, making it the de facto stronger sex. Physically, female bonobos are no stronger than males so, like beavers forever repairing their dams, they need to put continuous effort into staying on top. But beyond this truly remarkable achievement, the bonobo's political system is considerably less fluid than that of the chimpanzee. Again, this is because the most critical coalitions, those between mother and son, are unalterable. Bonobos lack the ever-shifting, opportunistic alliances capable of prying open a system. They are better described as tolerant than egalitarian.

Democracy is an *active* process: it takes effort to reduce inequality. That the more dominance-oriented, more aggressive of our closest relatives best demonstrates the tendencies upon which democracy ultimately rests is not surprising if we look at democracy as born from violence, as it most certainly is in human history. It is something we fight for: *liberté*, *égalité*, and *fraternité*. It has never been handed to us for free; it has always been wrested from the powerful. The irony is that we probably never would have reached this point, never would have evolved the necessary solidarity at the base, had we not been such hierarchical animals to begin with.

THREE

# SEX

## Kama Sutra Primates

This unusual and highly successful species spends a great deal of time examining his higher motives and an equal amount of time ignoring his fundamental ones. He is proud that he has the biggest brain of all the primates, but tries to conceal the fact that he also has the biggest penis.

—DESMOND MORRIS

In kinship with our insouciant, fun-loving, nonreading relatives the delightful cousins Bonobo. *May Life be thanked for them.*

—ALICE WALKER

A zoo keeper used to working with chimpanzees was introduced to bonobos and accepted a kiss from one of his new primate friends. In chimpanzees, a kiss is friendly rather than sexual. Was he taken aback when he felt the bonobo's tongue in his mouth!

The tongue-kiss—which the English language blames on the French—is an act of total trust: the tongue is one of our most sensitive organs, and the mouth is the body cavity that can do it the quickest harm. The act permits us to savor another. But at the same time we exchange saliva, bacteria, viruses, and food. Yes, food. In modern times, we may think of teenagers swapping chewing gum, but the mouth-kiss is thought to originate from maternal feeding of the young. And mother apes do indeed pass morsels of chewed food on to their offspring: from stuck-out lower lip into opened baby mouth. This is, of course, where the tongue comes in.

The French kiss is the bonobo's most recognizable, humanlike erotic act. Whenever I show an undergraduate class a film of my bonobos, the students get very quiet. They will watch all sorts of sexual intercourse, but invariably the deepest impression is made by a video clip of two juvenile males tongue-kissing. Even though one can never be sure exactly what goes on, it looks so ardent, so deep, with wide-open mouths placed on each other, that it takes my students by surprise. No Hollywood actor can match the zeal these juvenile apes put into the act. And the funny thing is that they will proceed seamlessly to a mock fight or burst into a playful chase. For bonobos, erotic contact mixes freely with everything else they do. They can move quickly from eating to sex, from sex to play, from grooming to a kiss, and so on. In fact, I have seen females continue collecting food while being mounted by a male. Bonobos take sex seriously, but never to the same degree as a lecture hall full of college students.

We humans set sex apart from our social lives, or at least we try to, but in bonobo society the two are fully intertwined. It's a human irony that all that our fig leaves seem to accomplish is insatiable sexual curiosity.

## PENIS ENVY

I sometimes get the impression that half the spam I receive on my computer has to do with the enlargement of a body part that men keep hidden most of the time. The male preoccupation with the size and turgidity of their manhood is an old source of income for snake-oil salesmen as well as the butt of endless jokes. From worship objects in ancient Greece and Rome to the phallic symbols that Sigmund Freud, chewing on his fat cigar, saw everywhere, the penis has long been said to have a mind of its own.

It was hardly surprising that Desmond Morris, when he shook us up in the sixties with unrestrained parallels between hairy and naked apes, chose to call attention to the size of the human penis, describing man as the sexiest primate to walk the earth. It was a brilliant move, designed to soften the blow he was delivering to our ego. There is nothing men want to hear more than that they are the champions in the one domain that counts. So little was known about the bonobo in those days that Morris may be forgiven for presenting us as the sexual title-holders. But even in this domain, we are not. It would be a challenge to actually measure the erection of an awake, aroused ape, but the bonobo's certainly makes most men look undersized. Even more so if we correct for the bonobo's smaller body size.

The bonobo's penis is thinner, though, and fully retractable, making an erection all the more eye-catching, especially if the male flicks his penis up and down, as they often do. Perhaps even more remarkable than the ability to "wave" his penis is that the bonobo's testicles are many times human size. Equally true for the chimpanzee, this is thought to relate to the amount of sperm needed for successful fertilization of females mating with several males. If the male bonobo genitalia strike us as well-developed, we will be even more impressed with the female's, because both chimps and bonobos carry enormous swellings. These are not the slightly swollen labia seen, if one peers carefully, in gorilla and orangutan females. No, they are soccer-ball-sized balloons on a female's behind that allow her to flash a brightly pink signal to all males in the neighborhood, telling them she's ready for action.

These swellings are made up of labia and clitoris. The bonobo clitoris is more prominent than in humans and chimpanzees. In young females it sticks out frontally like a little finger, while at a later age the clitoris will be embedded in the surrounding swelling tissue. Given this anatomy, it's not surprising that female bonobos prefer frontal copulation. Unfortunately for them, males rather seem to prefer the more ancient pattern of mating from behind. Female bonobos often invite males by lying on their backs, legs apart, or will quickly turn to this position if a male happens to start out differently.

As can be gathered from their remarks, zoo visitors are shocked by the eye-catching ape genitals. The most memorable reaction came from a woman who exclaimed, "Oh, my gosh, is that a head that I see?" Male apes are anything but confused: nothing is more exciting for them than a female with a voluminous pink behind. Personally, I'm so used to these striking features that they seem neither weird nor ugly, although the word "cumbersome" does come to mind. Fully swollen female apes cannot sit down normally; they awkwardly shift their weight between one hip and the other. They learn to deal with these appendices during adolescence, when they grow larger with every menstrual cycle. The swelling tissue is fragile, bleeding on the slightest occasion (but also healing quickly). On a par with human cultural inventions like foot binding and stiletto heels, it seems a heavy price for being attractive.

The bonobo's clitoris begs attention since its smaller human equivalent is at the center of such heated debate. What is a clitoris good for? Do we really need one? Theories range from how this tiny organ is totally useless, like a man's nipples, to how it's a source of pleasure that probably serves partner bonding. The first view assumes that women don't need to look for sex as long as they accept it when it knocks on their door. It calls the clitoris a "glorious accident" of evolution. In the second view, the clitoris evolved to enable orgasmic experiences so as to make sex a pleasant, addictive affair. The assumption here is that of an active female sexuality, one that seeks until it finds what it likes. These opposite views also align with opposing ideologies about the place of women in society.

Reproduction is too important to be left to chance. Every biologist expects both sexes, not just males, to be active in the choice of mates. We know that animals explore all options. In one amusing case, scientists hoped to control a population of red-winged blackbirds by vasectomizing the males.

They thought that nesting pairs with sterile males would produce sterile clutches. To their dismay, the majority of clutches were fertile, meaning that the females must have been "seeing" intact males on the side. The animal kingdom is full of sexually enterprising females who are shopping around, and human society is surely no exception. This is not usually picked up by surveys—a notoriously unreliable way of measuring behavior.

Surveys grossly underestimate women's sex lives: everybody, especially women, is reluctant to reveal the truth. We know this, because there *is* a way of getting women to talk. Hook up college students to a fake lie-detector machine, and young women report almost twice as many sex partners as women feeling no such pressure. In fact, they report as many partners as their male counterparts. So men and women may be far more similar than sex surveys have made us believe.

Since for males reproduction is a quicker affair than for females, it is often argued that the sexes should differ substantially in sexual tendencies. But not all sex is about making babies, neither in our own nor in many other species. What about pleasure and relaxation, what about oneness and bonding, and what about what my bonobos do every day: sex to iron out wrinkled relationships? Considering these other uses, the Victorian argument that sex is the provenance of men and a chore for women is based on rather narrow assumptions. If sex commonly expresses love, trust, and closeness, one would expect it to be at least equally a female domain. The French, who wisely stayed as far away from Queen Victoria as the Channel permitted, possess a wonderful range of expressions for these alternatives. Makeup sex is referred to as *la réconciliation sur l'oreiller* (reconciliation on the pillow), and the capacity of sex to set the mind at ease is rudely hinted at by describing a cantankerous woman as sexually deprived: a *mal baisée.*

Sex and sexual desire are supposed to go underground when the workday starts. A sharp boundary between the social and the sexual is universally human. Not, however, that it's perfectly maintained. In the old days, housemaids commonly were asked to provide services other than cooking and cleaning, and in modern society, romance often blossoms at the office, where there is plenty of sexual innuendo and harassment. On Wall Street, brokers are known to celebrate birthdays with strippers. But whatever the exceptions, as a rule, the social and sexual domains are kept separate.

We desperately need this boundary as our societies are constructed around family units involving paternal care as well as the maternal care natural to all mammals. Every human society has nuclear families, whereas apes have none. For chimps, the only time at which sex needs to be removed from the public domain is when a male and female worry about jealousy from higher-ups. They will arrange a rendezvous behind bushes, or travel away from the rest of the community, in a pattern that may be at the root of our desire for privacy. If sex is a source of tension, one way to keep the peace is by limiting its visibility. Humans take this even further, hiding not only the act itself but also covering up any arousing or arousable body parts.

None or very little of this occurs in bonobos. This is why they're often described as sexually liberated. But in fact, if privacy and repression are not an issue, where does liberation come in? They simply have no shame, no modesty, no inhibitions other than the wish to avoid trouble with rivals. When two bonobos couple, the young sometimes jump on top of them to take a peek at the details. Or another adult may move in and press her swelling against one of them to take part in the fun. Sexuality is more often shared than contested. A female may lie on her back masturbating in the open, and no one will blink an eye. She moves her fingers rapidly up and down through her vulva, but can also assign a foot to the job, keeping her hands free to groom her infant. Bonobos are great multitaskers.

Apart from claiming that penis size sets us apart from other primates, Desmond Morris argued that orgasm was exclusively human. Anyone who sees female bonobos engage in an intense sexual mount, known as GG-rubbing, however, will have trouble believing this claim. The females bare their teeth in a grin, uttering excited squeals as they frantically rub their clitorises together. Females also routinely masturbate, an activity that does not make sense unless they're getting something out of it. We know from laboratory experiments that we are not the only species in which females experience an increased heart rate and rapid contractions of the uterus at the climax of sexual intercourse. Macaque monkeys meet Masters and Johnson's criteria for human orgasm. No one has tried this same study on bonobos, but there can be little doubt that they would pass the test.

Not everyone is open to this possibility, though. One of the most curious academic meetings I ever attended was on the topic of sex. It was organized by postmodern anthropologists, who believe that reality is made up of words, that it cannot be separated from our narratives. I was one of only a handful of scientists at the meeting, and scientists by definition trust facts more than words. One can see how such a meeting would not go too well. Things came to a head when one of the postmodernists claimed that if a human language lacks a word for "orgasm," the people speaking this language cannot experience sexual climax. The scientists were taken aback. People all over the world have the same genitals and the same physiology, so how could their experiences be radically different? And what would this tell us about other animals? Would not the implication be that they feel nothing? Exasperated by the idea of sexual pleasure as a linguistic achievement, we began circulating little notes with naughty questions such as: Without a word for "oxygen," can people breathe?

Morris's final claim for human uniqueness ascribes to us the mating position thought to prove the sensibility of civilized people. Not only is the "missionary" position assumed to be restricted to our species, it is seen as a cultural advance. But considering the millions of years of sexual evolution behind us, I consider attempts to set human sexuality apart from that of other animals pretty hopeless. The hormones urging us to engage in sex, and the anatomical features making these peculiar acrobatics both feasible and pleasurable, are biologically dictated. They are far from unique: the way we do it is not that different from the way horses do it, or even the way guppies do it. From our frontally oriented genitals it's obvious that natural selection has favored the missionary position: we are anatomically designed to mate this way.

The same scientists who came up with the bonobo's unusual name wanted to explain their matings, but this topic was unmentionable at the time. Eduard Tratz and Heinz Heck had to resort to Latin, saying that chimpanzees mate *more canum* (like dogs) and bonobos *more hominum* (like people). Bonobos adopt the missionary position with great ease, as they do so many others. They know all positions of the Kama Sutra and even some positions that are beyond our imagination (such as both partners hanging upside down by their feet). The face-to-face position is special, though, in that it's both common and permits emotional exchange. Detailed video analysis shows that bonobos monitor the faces and sounds of their partners, regulating the speed of thrusting or rubbing to the response it elicits. If the partner fails to engage in eye contact, or otherwise seems lukewarm, the two split. Bonobos seem exquisitely attuned to their partners' experiences.

Bonobos not only have sex in a variety of positions but also in virtually all partner combinations. They disprove the notion that sex is intended solely for procreation. I estimate that three-quarters of their sexual activity has nothing to do with reproduction, at least not directly: it frequently involves members of the same sex or takes place during the infertile portion of a female's cycle. And then there are the many erotic patterns that do nothing for reproduction, including not only French kissing but also fellatio and massage of another's genitals, often seen among males. One male, with back straight and legs apart, will present his erect penis to another, who loosely closes his hand around the shaft, making caressing up-and-down movements.

The male equivalent of GG-rubbing is known as "rump-rump," a contact in which two males, standing on all fours, briefly rub their rumps and scrota together. It's a low-intensity greeting in which both partners face in opposite directions. Mutual penis rubbing, in contrast, resembles a heterosexual mount, with one male on his back and the other thrusting on him.

Because both males have erections, their penises rub together. I have never seen ejaculations during sex between males or attempts at anal penetration. All of this behavior has been seen in both captive and wild bonobos, except for so-called "penis-fencing," known only from studies in the natural habitat: two males hang face-to-face from a branch while rubbing their penises together as if crossing swords.

This richness of sexual behavior is truly amazing, yet it's been a mixed blessing for the bonobo's public profile. Some authors and scientists are so ill at ease that they talk in riddles. I have heard speakers call bonobos "very affectionate" while describing behavior that would be X-rated in any movie theater. Americans, in particular, avoid calling sex by its proper name. It's like listening to a gathering of bakers who have decided to drop the word "bread" from their vocabulary, making for incredibly circumlocutory exchanges. The sexiness of bonobos is often downplayed by counting only copulations between adults of the opposite sex. But this really leaves out most of what is going on in their daily lives. It is a curious omission, given that the "sex" label normally refers to any deliberate contact involving the genitals, including petting and oral stimulation, regardless of who does it to whom (when President Bill Clinton tried to define sex more narrowly, he was corrected by the courts). In a broader sense, sex also covers kissing or showing off one's body in a suggestive manner, which is precisely why "Elvis the Pelvis" was abhorred by the parents of the 1950s. Personally, I'm all for calling a spade a spade—euphemisms born from squeamishness have no place in scientific discourse.

If I leave the impression that the bonobo is a pathologically oversexed animal, I must add that the bonobo's sexual activity is remarkably casual, far more so than our own. Like people, they engage in sex only occasionally, not continuously. Many of the contacts are not carried through to the point of climax—the partners merely fondle each other. Even the average copulation is incredibly quick by human standards: fourteen seconds. Instead of an endless orgy, what we see is a social life salted and peppered by brief moments of sexual intimacy. To have such a lusty close relative, however, has implications for how we view our own sexuality.

## BI BONOBOS

Do bonobos really need all this sex? Do we humans need it? Why bother with it at all? This may seem a strange question—as if we have a choice!—but instead of taking sex for granted, biologists wonder where it came from, what it's good for, and if there are perhaps better ways to reproduce. Why don't we just clone ourselves? Cloning has the advantage of replicating genetic designs that have worked well in the past, such as you and me (being alive after so many years is a major achievement), without mixing in the imperfections of somebody else's genes.

Imagine the brave new world we would inhabit, full of genderless, identical-looking individuals. No more gossip about who loves whom, who divorces whom, or who cheats on whom. No unwanted pregnancies, no silly magazine articles about how to impress your date, and no sins of the flesh but also no infatuation, no romantic movies, and no pop star sex symbols. It might be more efficient, but it would also be the most boring place one could imagine.

Fortunately, the drawbacks of sexual reproduction are more than made up for by the benefits. This is nicely demonstrated by animals who employ both methods of copying. Pick up one of those aphids found on houseplants, for example, and put her under a microscope. Within her translucent belly one sees a set of tiny daughters, all identical to mom. Most of the time, aphids just clone themselves. When times get tough, however, such as in fall and winter, this method is not good enough. Cloning doesn't allow them to get rid of random genetic mutations, most of which cause problems. Errors will accumulate until the whole population drowns in them. So aphids switch to sexual reproduction, which brings gene-mixing. Sexually produced offspring are more robust in the same way that, for example, a mixed-breed dog or cat is generally healthier than a purebred. After many generations inbreeding resembles cloning, resulting in ever more genetic defects.

The vigor of the so-called "wild type"—the product of sexual reshuffling of the gene deck—is well known. It handles disease better, for instance, since it's able to keep up with the continuous evolution of parasites. Bacteria need only nine years to go through the approximately 250,000 generations our lineage has gone through since we split off from bonobos and chimpanzees. The quick generational turnover of parasites forces host animals to change their defenses. Just to fight off parasites, our immune systems need to constantly stay on the move. Biologists call this the Red Queen Hypothesis, after the Red Queen in *Alice's Adventures in Wonderland*, who told Alice, "It takes all the running *you* can do to stay in the same place!" For man and animals alike, the running is done through sexual reproduction.

But this only explains why sex exists, not why we engage in it so often. Would we not reproduce quite nicely with only a fraction of it? This is certainly what the Catholic Church has in mind when it claims sex is intended only for reproduction. But doesn't the pleasurable side of sex challenge this view? If its sole aim were reproduction, surely sex would not need to be such fun. We would rather be looking at it the way children look at vegetables: recommended yet unexciting. This is, of course, not exactly what Mother Nature had in store for us. Fed by myriad nerve endings in places known as erogenous zones (eight thousand in the little clitoris alone) wired directly to the brain's joy centers, sexual desire and satisfaction have been built into our bodies. Pleasure-seeking is the number one reason why people have more sex than is necessary for reproduction.

The discovery that one of our closest relatives has genitalia that look at least as well-developed as ours, and engages in even more "unnecessary" sex than we do, makes sexiness a majority trait in the trio of close relatives we are considering. Chimps are the exception. Their sex life is poor compared to that of people and bonobos, and not just in the wild, but also at the zoo.

Comparing captive chimps and bonobos that have equal amounts of space, get the same food, and have the same number of partners, bonobos initiate sex on average once every one-and-a-half hours and with a far greater diversity of partners than do chimps, who have sex only once every seven hours. Thus, even in the same environment, bonobos are far more sexual.

But none of this answers the question at hand—why are people and bonobos such sexual hedonists? Why are we endowed with sexual appetites beyond those needed to fertilize the occasional egg and beyond the partners who make this possible? Readers may object that their taste in sex partners is less varied than this, but I'm thinking about us *as a species.* Some people are heterosexual, some are gay, and some like a variety of partners. Moreover, these classifications seem arbitrary. The American sex research pioneer Alfred Kinsey placed human sexual preferences on a smooth continuum, opining that the world is not divided into sheep and goats and that our usual distinctions are not nature's but society's doing.

Kinsey's view is supported by cross-cultural studies, which indicate incredible variation in attitudes toward sex. Homosexuality is freely expressed, even encouraged, in some cultures. The ancient Greeks come to mind, but there is also the Aranda tribe of Australia, among whom an older man lives sexually with a boy until the former is ready to marry a woman, and in which women rub each other's clitorises for pleasure. In the Keraki of New Guinea, intercourse with men is part of every boy's puberty rites, and there are other cultures in which boys fellate older men in order to ingest sperm, supposedly to gain masculine strength. Contrast this with cultures that surround homosexuality with fear and taboos, especially in men, who underline their manhood by stressing their heterosexuality. No heterosexual male wants to be mistaken for a homosexual. Intolerance forces everyone to carve up their sexuality and chose among its parts, giving the impression of discrete types even though, underneath, a wide range of preferences may exist, including individuals with no preference at all.

I'm stressing this cultural overlay to explain why the common evolutionary question of how homosexuality could have arisen may be beside the point. Since homosexuals do not reproduce, so the argument goes, they should have gone extinct a long time ago. But this is a puzzle only if we buy into modern labeling practices. What if declared sexual preferences are mere approximations? What if we have been brainwashed into either/or thinking? And what about the premise that homosexuals fail to reproduce? Is this really the case? They are certainly capable of reproduction, and in modern society many are married during some stage of their lives. Plenty of gay couples are raising families in our world. The extinction argument also assumes a genetic gulf between homosexuals and heterosexuals. It is true that sexual preferences appear constitutional—meaning that we are born with them or at least develop them early in life—but rumors of "gay genes" notwithstanding, there is as yet no evidence for a systematic genetic difference between homosexuals and heterosexuals.

Let us step away from the sexual domain and just speak of attraction to one's own sex, assuming that this attraction exists to some degree in all of us. We easily bond with individuals who are like ourselves, so this part is not hard to follow. Inasmuch as same-sex attraction does not prevent attraction to the opposite sex, its evolution would have met with no obstacles. Now, add to this the idea that there exists a gray area between social attraction on the one hand and sexual attraction on the other. That is, same-sex attraction may have sexual undertones that surface only under certain circumstances. If the opposite sex is out of sight for a long time, for instance, such as at boarding schools, in prisons, nunneries, or on ships, same-sex bonding often turns sexual, which might not have happened otherwise. And when inhibitions disappear, such as when men drink too much, all of a sudden they hang around each others' necks. The idea that attractions that are not consciously sexual can still have a sexual side is, of course, far from new: Freud impressed this upon us long ago. We are so afraid of sex that we try to squeeze it into a little box with a lid on it, but it slips out all the time, mixing with a multitude of other tendencies.

Attraction to our own gender is not an evolutionary problem inasmuch as it does not conflict with reproduction. Let's further assume that this attraction is highly variable, with its social side winning out in the majority of individuals and its sexual side in a minority. This minority is small. Kinsey's estimate of 10 percent homosexuals in the population was a gross exaggeration. Most recent surveys report less than half this figure. Within this minority exists an even tinier group with a same-sex attraction so strong that it precludes heterosexual sex, hence reproduction. The largest randomized study of sexual behavior to date, carried out in the 1990s in the United States and Great Britain, puts the number of exclusive homosexuals below 1 percent. Only if this mini-minority were to carry genes that no one else has, would there be a puzzle—how would these genes be passed on? But, as said, there is no solid evidence for such genes. Besides, the 99 percent of the population with the ability to reproduce seems to have no trouble passing on the same-sex attraction of which homosexuality seems an outgrowth.

Rather than a "lifestyle choice," as some conservatives wishfully label it, this outgrowth comes naturally to certain individuals. It's part of who they are. In some cultures they are free to pursue it, whereas in others they need to hide it. Since there are no cultureless people, it's impossible to know how our sexuality would look in the absence of these influences. Pristine human nature is like the Holy Grail: eternally looked for, never found.

But we do have the bonobo. This ape is instructive since it knows no sexual prohibitions and few inhibitions. Bonobos demonstrate a rich sexuality in the absence of the cultural overlays that we create. This is not to say that bonobos are furry people: they are clearly a separate and different species. On Kinsey's 0–6 scale from hetero- to homosexual, humans may be predominantly at the heterosexual end but bonobos seem totally "bi," or a perfect Kinsey 3. They are literally pansexual—a most fortuitous label, given their genus name. As far as we know, there are no exclusively hetero- or homosexual bonobos: all of them engage in sex with virtually all kinds of partners. When this news about one of our closest relatives broke, I was drawn into a discussion on a gay Internet site where some argued that this meant that homosexuality was natural, whereas others complained that it made it look primitive. With "natural" sounding good and "primitive" sounding bad, the question was whether the gay community should or shouldn't be happy with bonobos. I didn't really have an answer: the bonobo is here, whether people like it or not. But I did suggest they take "primitive" to mean what it means in biology—namely, the more ancestral form. In this sense, heterosexuality is obviously more primitive than homosexuality: in the beginning there was sexual reproduction, leading to two sexes and a sex drive. Additional applications of this drive must have come later, including same-sex sexual relations.

Same-sex sex is certainly not limited to humans and bonobos. Monkeys mount members of their own sex to demonstrate dominance and are also known to present their behinds in appeasement. In macaques, females consort with each other like a heterosexual pair, one female regularly climbing on top of the other. Example after example of homosexual sex has been documented in the animal kingdom, from mounting bull elephants and necking giraffes to the greeting ceremonies of swans and mutual caresses among whales. But even if some animals go through periods in which such behavior is common, I shy away from the term "homosexual" and its implication of a predominant orientation. Exclusive same-sex orientations are rare or absent in the animal kingdom. Bonobos are sometimes presented as gay animals, leading to a Bonobo Bar in almost every cosmopolitan city. True, bonobos have frequent homosexual sex, if we use this term just to denote the act. Females do it with each other all the time, and in fact GG-rubbing is the political cement of their society. It clearly is part of female bonding. Males regularly engage in sexual behavior with each other, too, although less intensely than females. But none of this is enough to make bonobos gay. I know of no bonobos who restrict sex to members of their own gender. Instead, they are fully promiscuous and bisexual.

The most significant point about bonobo sex is how utterly casual it is, and how well integrated with social life. We use our hands in greetings, such as handshakes and pats on the shoulder, whereas bonobos offer genital handshakes. Let me describe a scene at the Wild Animal Park, northeast of San Diego, when the keepers and I provided the bonobos with a meal for them to share while a camera team recorded their table manners for a popular science program. We filmed them in a spacious grassy enclosure with palm trees. Even though a fully grown muscular male named Akili was present, the group was ruled by Loretta, then twenty-one years old. The apes did exactly what they were supposed to do: resolve tensions over food with sex.

When a large bundle of ginger leaves—a favorite food—was thrown in front of the group, Loretta immediately seized it. After a while, she allowed Akili to eat some, but a younger adult female, Lenore, hesitated to join in. This wasn't because of Loretta, but because Lenore and Akili for some reason didn't get along. The caretaker told me that this had become a persistent problem. Lenore kept looking at Akili, avoiding his every move. Lenore presented from a distance a number of times. When Akili failed to respond, she approached him and rubbed her genital swelling against his shoulder, which he accepted. After this, she was allowed to join the group, and all of them peacefully ate together, though Loretta still firmly controlled the food.

The group also included an adolescent, Marilyn, who had something else on her mind. She was enamored of Akili and followed him around, sexually inviting him many times. Marilyn played for a while in the pool, manually stimulating her genitals while dipping her lips into the water. After having excited herself in this manner, she pulled at Akili's arm and led him by the hand to the water for copulation. Akili obliged multiple times, but was clearly torn between Marilyn and the food bonanza. Why the sex had to take place in knee-deep water was unclear to me: perhaps Marilyn had developed a water fetish. Sexual idiosyncrasies are not unusual in bonobos.

In the meantime, Loretta showed great interest in Lenore's baby. Whenever the infant came close, she would briefly stimulate its genitals with a finger, one time followed by a belly-to-belly embrace in which Loretta made a series of malelike pelvic thrusts against the infant. At one point, the mother manually stimulated Loretta's genitals after which she pushed her offspring toward Loretta as if urging her to hold it.

In this short stretch of time we saw bonobos use sex for sex (Akili and Marilyn), for appeasement (Lenore and Akili), and as a sign of affection (Loretta with the infant). We usually associate sex with reproduction and desire, but in the bonobo it fulfills all sorts of other needs. Gratification is by no means always the goal, and reproduction is only one of its functions.

## OF LADIES AND TRAMPS

Bonobo females sport genital swellings even when they're not fertile, such as during pregnancy or lactation. This is not true for chimps. It has been calculated that female chimps are swollen less than 5 percent of their adult life, whereas bonobo females are in this state close to 50 percent. Moreover, except for a dip when females have their period, bonobo sex keeps going throughout the cycle. For a primate with swellings, this is perplexing. What are those grotesque balloons good for except announcing fertility?

With sex and swellings being largely disconnected from fertility, a bonobo male would need to be an Einstein to figure out which youngsters he may have fathered. Not that apes are aware of the link between sex and reproduction—only we are—but it is quite common for male animals to favor offspring of females they have had sex with, thus in effect caring for and protecting their own progeny. Bonobos have too much sex with too many partners to make such distinctions, however. If one had to design a social system in which fatherhood remained obscure, one could scarcely do a better job than Mother Nature did for the bonobo. We now believe that this may, in fact, be the whole point: females gain from luring males into sexual relations. Again, no conscious intent is implied, only the misrepresentation of fertility. At first, this idea is bewildering. Although paternity is never as sure as maternity, isn't our species doing rather well with a high confidence in fatherhood? Men are quite a bit more certain about this issue than male animals with unlimited promiscuity. What could be wrong with males knowing who their offspring are? The reason is infanticide: the killing of infants by males.

I happened to be present at the historic meeting in Bangalore, in southern India, at which Yukimaru Sugiyama, a well-known Japanese primatologist, first reported how male langur monkeys take over a harem of females by ousting the old leader and then customarily kill all the infants. They snatch them from their mothers' bellies, impaling them with their canine teeth. The meeting took place in 1979, and at the time no one realized that history was being made, that one of the most provocative evolutionary hypotheses of our time was being born.

Sugiyama's presentation met with a deafening silence, followed by dubious praise from the session chair for these intriguing examples of what he called "behavioral pathology." These were the chairman's words, not the speaker's. The idea that animals kill their own kind, and not just accidentally, was both incomprehensible and repulsive.

Sugiyama's discovery and his speculation that infanticide might help male reproduction was ignored for a decade. But then other reports surfaced, first in other primates and eventually in many other animals: from bears and prairie dogs to dolphins and birds. When male lions take over a pride, for instance, the lionesses put all their energy into an attempt to stop them from hurting the cubs, but usually to no avail. The king of animals jumps on the helpless cubs and shakes them with a neck bite, killing them instantly, without eating them. It looks totally deliberate. There was utter disbelief in the scientific community that the very same theories that talk of survival and reproduction might apply to the annihilation of innocent newborns

But this is exactly what was being proposed. When a male takes over a group, he not only chases off the previous leader, but also removes this male's last reproductive efforts. This way, the females will resume cycling sooner, thus helping the new male's own reproduction. Sarah Blaffer Hrdy is an American anthropologist who became the architect of this idea, as well as drew attention to examples of human infanticide. It's well established, for example, that children are more at risk of abuse by stepfathers than by biological fathers, which seems to fit a connection with male reproduction. The Bible describes the Pharaoh ordering children killed at birth and, most famous, King Herod who "sent forth, and slew all the children that were in Bethlehem, and in all the coasts thereof, from two years old and under" (Matthew 2:16). The anthropological record shows that after war, it's quite common for children of captured women to be killed. There is every reason to include our own species in discussions of male infanticide.

Infanticide is regarded as a key factor in social evolution, pitting male against male and male against female. Females stand to gain nothing: the loss of an infant is always disastrous. Hrdy theorized about female defenses. Of course, females do what they can to defend themselves and their offspring, but given the larger size of males and their special weaponry (such as large canine teeth), this is often futile. The next best thing is to confuse paternity. When outside males take over, such as in lions and langur monkeys, a newcomer can be 100 percent certain that he didn't father any of the young he sees around. But if a male already lives in the group where he meets a familiar female with an infant, the situation is different. The infant may well be his, meaning that killing it would reduce the male's ability to pass on his genes. From an evolutionary perspective, nothing could be worse for a male than to eliminate his own progeny. It's assumed, therefore, that nature has provided males with a rule of thumb to attack only infants of mothers with whom they have had no recent sex. This may seem foolproof for the males, but it opens the door for a brilliant female counterstrategy. By accepting the advances of many males, a female can buffer herself against infanticide because none of her mates can discard the possibility that her infant is his. In other words, it pays to sleep around.

So there you have it: the possible reason why bonobos have lots of sex and no infanticide. The latter has never been observed, neither in the field nor in captivity. Males have been seen to charge females with small infants, but the massive defense against such behavior suggests formidable opposition to infanticide. The bonobo is a true exception among the apes, because infanticide is well documented in gorillas and chimpanzees, not to mention humans. A large male chimp in Budongo Forest in Uganda was discovered holding a partly eaten dead infant of his own kind. Other males were nearby, and the carcass was passed back and forth between them. Dian Fossey, of *Gorillas in the Mist* fame, saw a lone silverback gorilla enter a troop with a violent charge. A female who had given birth the night before countered his charge, standing upright beating her chest. The newborn on her exposed belly was struck at once by the male: it died with a wail.

Naturally, we find infanticide revolting. One field-worker couldn't resist stepping in as male chimps surrounded a female who crawled on the ground, trying to hide her infant, fervently pantgrunting to stave off an attack. The field-worker forgot her professional obligation not to interfere and confronted the males with a large stick. This may not have been the smartest thing to do since male chimpanzees sometimes kill people, but the scientist got away with it and managed to scare them off.

No wonder female chimps stay away from large gatherings of their species for years after having given birth. Isolation may be their primary strategy for prevention of infanticide. They resume swellings only toward the end of the nursing period, after three to four years. Until this time, they have nothing to offer males looking for sex and no effective way to change the mind of males in an aggressive mood. Female chimps spend a large part of their lives traveling alone with their dependent offspring. Bonobo females, in contrast, rejoin their group right away after having given birth and copulate within months. They have little to fear. Bonobo males are not in any position to know which offspring are theirs. And since female bonobos tend to be dominant, attacking their offspring would be a risky business.

Free love born of self-protection? "That's why the lady is a tramp," Frank Sinatra would tell us. "She loves the free, fresh wind in her hair / Life without care," he sings. And indeed the insouciant existence of female bonobos contrasts with the dark cloud under which many other female animals live.

It's hard to overestimate the premium evolution must have placed on putting an end to infanticide. Bonobo females battle for a cause—the most urgent one imaginable for their gender—with all weapons at their disposal, both sexual and aggressive. They seem triumphant.

These theories fail to explain the bonobo's varied sexuality, though. The way I imagine the origin of such behavior is that once evolution had turned bonobos into heterosexual party animals, sex simply spilled over into other domains, such as same-sex bonding and conflict resolution. The species became sexualized through and through, as probably also reflected in their physiology. Neuroscientists have discovered some intriguing things about oxytocin, a common hormone in mammals. Oxytocin stimulates uterine contractions (it's regularly given to women in labor) and lactation, but what's less known is that it also reduces aggression. Inject a male rat with this hormone and he's far less likely to attack pups. Even more interesting is that the synthesis of this hormone in the male brain spikes after sexual activity. In other words, sex produces a touchy-feely hormone that, in turn, instills a peaceful attitude. Biologically, this might explain why human societies in which physical affection is common and sexual tolerance high are generally less violent than societies not so inclined. People in these societies may have higher oxytocin levels. No one has measured oxytocin in bonobos, but I bet they're brimming with the stuff.

John Lennon and Yoko Ono may well have been right when they staged a weeklong "bed-in" against the Vietnam War at the Amsterdam Hilton: love brings peace.

## CHASTITY BELTS

If bonobo females are a rare success story within MAI, the imaginary organization Mothers Against Infanticide, this raises questions about ourselves. Are women active in MAI, too?

Instead of following the bonobo model, our species has chosen its own way. Two elements women share with bonobos are that their ovulation is hidden from immediate detection and that they have sex throughout their cycle. But here the similarities end. Where are our genital swellings, and where is the sex at the drop of a hat?

To start with swellings: scientists have wondered why we lost them and have even speculated that our fleshy buttocks have taken their place. And not merely because buttocks occupy the same physical location, but because they, too, enhance sex appeal. This is an odd thought, however, because shouldn't this have led to different buttocks in women than in men? Connoisseurs that we all are, we have no trouble telling the behinds of men from those of women, even underneath a layer of clothes, yet it cannot be denied that they are more similar than different. This does us no good as a sexual signal. It seems far more likely that we never had any genital swellings to begin with. Swellings likely evolved after the human-ape split and only in the *Pan* lineage, because they are found in no other apes.

When women began stretching their sexual receptivity, surpassing even the bonobo's tenfold increase beyond and above the chimpanzee's, they didn't need to lengthen any swelling phase. Rather than advertise untruthfully, our method was to skip advertising altogether. Why didn't bonobos evolve this more convenient method, too? My guess is that once swellings existed and became a male fixation, the clock couldn't be turned back. Females with reduced swellings would have lost out to those better endowed. This is a familiar story in sexually selected traits, such as the ever-expanding peacock's tail. Competition about who is sexiest often leads to exaggerated signals.

The second way in which we deviate from bonobos is that sex among humans is more restrained. This is not always evident, because some societies know extraordinary freedom. The peoples of the Pacific were a case in point before the arrival of Westerners, who brought both Victorian values and venereal disease. In *The Sexual Life of Savages*, Bronislaw Malinowski depicted this region's cultures as having few taboos or inhibitions. In a rather bonoboesque comment, it has been said that for early Hawaiians "sex was a salve and a glue for the total society." The Hawaiians worshipped genitals in song and dance, pampering these body parts in their children. Breast milk was squirted into an infant's vagina and the labia molded together so that they would not separate. A little girl's clitoris was stretched and lengthened through oral stimulation. The penis received similar treatment so as to enhance its beauty and prepare it for sexual enjoyment later in life.

Unrestrained hedonism, however, is unlikely to have existed in any human culture. Some anthropologists, such as Margaret Mead, relying on informants rather than firsthand observation, have created a romantic fiction that is still with us today. But even the most sexually liberal cultures are not free from jealousy and violence in response to unfaithfulness. Universally, intercourse takes place in private, and the genital region tends to be hidden. That even the early Hawaiians knew chastity is suggested by their word for loincloth, *malo,* the most likely origin of which is *malu,* the Malayan word for "shame."

Most societies also limit sex to a few partners. Polygamy may be practiced and accepted, but in reality the vast majority of families in the world include only one man and one woman. The nuclear family is the hallmark of human social evolution. Given the exclusivity of our sexual contacts, we've opted for the opposite of the bonobo plan by actually enhancing a male's ability to tell which offspring are his. Until modern science came along, men could never be sure, but had a far better chance of guessing right than a bonobo does.

Natural selection shaped human behavior in response to pressures quite different from those encountered by the apes. Our ancestors had to adapt to an incredibly harsh environment. They left the protection of the jungle for the flat, dry savanna. Don't believe those killer-ape stories of Robert Ardrey and others about how our ancestors ruled the savanna as top predators. Our forebears were *prey.* They must have lived in constant fear of pack-hunting hyenas, ten different kinds of big cat, and other dangerous animals.

In this scary place, females with young were the most vulnerable. Unable to outrun predators, they could never have ventured far away from the forest without male protection. Perhaps bands of agile males defended the group and helped carry juveniles to safety during emergencies. This would never have worked, though, if we'd kept the chimp or bonobo social system. Promiscuous males are simply not good at commitment. Without hope to single out their progeny, they have little reason to invest in child care. To get the males involved, society would have had to change.

Human social organization is characterized by a unique combination of (1) male bonding, (2) female bonding, and (3) nuclear families. We share the first with chimpanzees, the second with bonobos, and the third is ours alone. It's no accident that people everywhere fall in love, are sexually jealous, know shame, seek privacy, look for father-figures in addition to mother-figures, and value stable partnerships. The intimate male-female relationship implied in all of this, which zoologists have dubbed a "pair-bond," is bred into our bones. I believe this is what sets us apart from the apes more than anything else. Even Malinowski's hedonistic "savages" were not without a tendency to form exclusive households in which both males and females cared for children. Our species' social order revolves around this model, which gave our ancestors a foundation for building cooperative societies to which both sexes contributed and in which both felt secure.

It has been speculated that the nuclear family arose originally from a male tendency to accompany females with whom they had mated so as to keep infanticidal rivals at bay. Such an arrangement could have been expanded to include paternal care. For example, the father could have helped his companion locate trees with ripe fruits, capture and share prey, or carry their young. He himself would have profited from her talent for precision tool-use (female apes are more skillful than males) and her gathering of nuts and berries. The female, in turn, may have begun offering sex so as to keep her protector from running off with every good-looking passerby. The more both parties put into this arrangement, the higher the stakes. Thus, it became increasingly important to the male that his mate's offspring were his and only his.

There is no free lunch in nature. If bonobo females paid for their arrangement with almost continuous swellings, women paid for theirs with diminished sexual freedom. And the motivation for male control only increased when our ancestors settled down from a nomadic existence and began to accumulate material goods. In addition to passing on genes to the next generation, they were now also inheriting wealth. The size difference between the sexes combined with excellent cooperation among males makes it likely that male dominance has always characterized our lineage, and so inheritance likely followed paternal lines. With every male trying to ensure that his life's savings ended up in the right hands—those of his own progeny—an obsession with virginity and chastity became inevitable. Patriarchy, as it's known, can be thought of simply as an outgrowth of male assistance with the rearing of offspring.

Many of the moral constraints we're used to—including those that would land bonobos in jail were they to live amongst us—are designed to uphold this particular social order. Our ancestors needed cooperative males who posed no threat to females and their young and who were ready to lend their mates a hand. This meant separation between public and private spheres and exclusive partnerships. We needed to rein in ancient promiscuous tendencies that must have lingered for a while, and linger still. The result was not only survival but increased population growth compared to apes. Female chimpanzees give birth only once every six years, whereas bonobos (who live in a more food-abundant environment) do so about every five years. The latter birthrate may be the best apes can manage, given that they nurse their young for four or five years and carry them for about the same time. Bonobo females sometimes give birth in such quick succession that they end up nursing two offspring at the same time. Having no strollers or sidewalks, a bonobo female may climb around the forest with an infant clinging underneath her belly and a juvenile jockey-riding on her back. This seems an unbearable burden. Bonobos have stretched the single-parent system to the limit.

Fatherly assistance allows earlier weaning, which explains why we, and not the apes, colonized the planet. But since males are willing to help out only with young likely to be theirs, the taming of female sexuality became their constant struggle. In recent times, we've seen efforts at male control in the extreme under the Taliban in Afghanistan. Their Department for the Preservation of Virtue and Prevention of Vice meted out public floggings for women who allowed their faces or ankles to show. But also in the West, there is no lack of rules regulating sex, always applied more strictly to women than men. It's our familiar double standard, for example, that leads health insurance companies to cover Viagra but not the morning-after pill. In every language, labels for adulterous women are far worse than those for similarly adventurous men. Where a woman is a "slut," a man is merely a "skirt chaser."

Human evolution has been curiously uncooperative, however, in upholding the reproductive purity of the family. Imagine extraterrestrial visitors digging up a chastity belt, trying to figure out what this thing was for. The iron or leather contraption fits around a woman's hips, reaching around the anus and vulva, leaving openings too small for sex but large enough for other functions. Fathers or husbands held the keys. One doesn't need to be a rocket scientist to understand why belts were better at putting men's minds at ease than moral codes. The female of our species is only moderately faithful. If faithfulness had been nature's goal, female sexual appetite would have been limited to the fertile period, and this phase would have been made externally detectable. Instead, nature has created a female sexuality that is almost impossible to control. The common argument that men are naturally polygamous and women naturally monogamous is as full of holes as Swiss cheese. What we see in reality is a mismatch between our social organization, which revolves around the nuclear family, and our sexuality.

Blood and DNA tests conducted in Western hospitals suggest that about one in fifty children is not sired by the father of record. In some studies, the mismatch is considerably greater. With all these "mama's babies but daddy's maybes" out there, no wonder it's a child's resemblance to the father that people most often stress. It's remarkable how often even mothers themselves say "just like daddy." We all know which of the two parents needs reassurance.

Few societies openly tolerate extramarital affairs, although a few do, such as the Barí Indians of Venezuela, who have a rather bonobolike system. The bonobolike part is that women have sex with multiple partners, and hence confuse paternity. The humanlike part is that this helps women secure male care. The Barí Indians believe that once a fetus is formed, usually by husband and wife, it needs to be nourished with semen, so both the husband and the woman's other partners contribute to its growth. (This sounds odd to modern ears, but scientific proof that eggs are fertilized by only one sperm cell dates back no further than the nineteenth century.) Once born, the baby is not considered the product of a single but of several biological fathers. Joint paternity has distinct advantages in cultures with high child mortality. It's difficult for a single father to adequately provide for his family. If several men feel an obligation, this helps child survival. Women in effect buy themselves child support by having sex with several men.

If the nuclear family doesn't always conform to the way Western biologists see it—a male helping his mate in return for her faithfulness—the basic idea is still the same: women seek as much protection and care as they can get, and men are drawn into this by sexual relations. Sometimes women consider brothers more reliable helpers than their mates might be, but the typical human pattern by far is a sex-for-food deal between a man and a woman with children attached.

## THE GREAT INSEMINATOR

In *Totem and Taboo*, Sigmund Freud imagined that our history began with what he called "Darwin's primal horde." A jealous, violent father kept all women for himself, driving away his sons as soon as they were grown. This triggered a revolt against his authority. The sons banded together to kill their father, after which they devoured him. They consumed him both literally and metaphorically—internalizing his strength and identity. During his reign they had detested him, but following his death they could finally acknowledge the love they had also felt. Thus, there was remorse, followed by adulation, and ultimately the emergence of the God concept. "At bottom," Freud concluded, "God is nothing other than an exalted father."

Religions tend to present sexual morals as God-given and in doing so harken back to the image of an ancestral alpha male that, according to Freud, has always kept a firm hold on our psyche. It's a fascinating thought that ancient patterns of sexual competition are perpetuated by religions without our realizing it. But these patterns have endured in real life as well. Anthropologists have provided us with ample evidence that powerful men control more women and produce more offspring. A staggering example comes from a recent genetic survey in Central Asian countries. The study concerned the Y-chromosome, which is found in males only. No less than 8 percent of Asian men possess virtually identical Y-chromosomes, suggesting a single forefather for all of them. This male had so many children that he now has an estimated sixteen million male descendants. Having determined that this great inseminator lived about a thousand years ago, scientists have settled on Genghis Khan as the most likely candidate. Khan, his sons, and grandsons controlled the largest empire in world history. Their armies slaughtered entire populations. Beautiful young women were not for the troops, however: they were brought to the Mongolian ruler himself.

The tendency of powerful males to claim a disproportionate piece of the reproductive pie remains. But raw male rivalry has been replaced with a system in which potentially every male has his own family, and the entire community sanctions and respects the bond with his mate. That this arrangement may have been with us for quite a while is suggested by both the size difference between men and women and, strangely enough, the dimensions of our testicles. There are two hundred primate species, and those in which a single male monopolizes multiple females always have males who are far larger than the females. Freud's primal horde sounds a lot like a gorilla harem, with the feared father twice the size of his many mates. Ironically, however, the more domineering the male of a species, the tinier his testicles. A gorilla's testes weigh extremely little compared to his overall body size. This makes sense, because no other male ever gets close to a ruling male's females. Since he's the only one doing the fertilizing, small amounts of sperm suffice.

Contrast this with the promiscuous chimps and bonobos, with many males vying for the same females. If a female mates with several males on the same day, the sperm cells of all of her lovers will be racing toward the egg. This is known as sperm competition. The male who dispatches the most and the healthiest sperm wins. In this instance, males don't need to be as large and heavily armed as the great harem holders of the animal kingdom, such as walruses, gorillas, deer, and lions. When animals rely on sperm competition, females are not much smaller than males. Female chimps weigh around 80 percent of what their male counterparts weigh, and for bonobos and humans the sexes are even slightly closer. All three species, but the last two in particular, show signs of reduced male competition. There is one major difference, however: both chimps and bonobos are far more promiscuous than we are. Our testicles reflect this: they are mere peanuts compared to our ape relatives' coconuts. Corrected for body size, chimp testes are about ten times larger than those of the human male. Bonobo testes have not been carefully measured, but they appear larger than those of chimps, although the bonobo's body is smaller. So in this regard, too, bonobos seem the champions.

Science has spent far more ink on the size of our brains than on the size of our balls. Yet in the larger scheme of animal behavior, genital comparisons are extremely instructive. They suggest that in humans are combined two elements not found together in any of our close relatives: a multi-male society with low sperm competition. The Genghis Khan story notwithstanding—

which mainly concerns competition over females outside a male's own group—our small testicles indicate that, for the most part, our male ancestors were not all going after the same females. Something must have kept them from rampant promiscuity. Something must have made them deviate fundamentally from the open mate competition among chimps and bonobos. This "something" is no doubt the nuclear family, or at least the existence of stable heterosexual pair-bonds. Our anatomy tells a story of romance and bonding between the sexes going back a long time, perhaps to the very beginnings of our lineage. This is supported by *Australopithecus* fossils of which the minor size difference between males and females hints at a monogamous society.

Despite this heritage, the effects of male dominance and privilege remain present in our societies, not only insofar as certain men claim more sexual partners than others, but also in the treatment of women. When males dominate, they have ways of extracting sex—"rape" in humans and "forced copulation" in animals. Let me add right away, though, that the fact that such behavior occurs doesn't mean it's biologically dictated. A recent book claiming that rape is natural caused incredible uproar mainly because it was seen as an attempt to justify this behavior. The idea originally derived from research on insects, in which some species have anatomical features—a sort of clamp—that help males force females into sex. Men obviously lack such features, and even though the underlying psychology of rape (such as violent predispositions or lack of empathy) may very well have a genetic side, to think that rape itself is encoded into us is like assuming that some people are born to burn down houses or write books. The human species is far too loosely programmed for such highly specific behavior to be genetic.

Whether we have humans or apes in mind, involuntary intercourse is better looked at as an *option* available to any male who desires a female and is able to control her. Bonobo males lack this option since the females are dominant and hence never resort to anything even remotely like it. Chimp males are different, however, and forcing females into sex is not beyond them. In captivity, this is rare, thanks to the efficiency of female alliances. I have seen males bluff and intimidate females unwilling to mate, but there is almost always a point at which other females come to her rescue and collectively interrupt the unwanted advances. In human societies, too, rape and sexual harassment are less common whenever women are surrounded by a support network of kin.

Wild female chimps, on the other hand, are vulnerable as they often travel alone. Males may get away from the tense atmosphere with other males by taking a swollen female on a "safari." They take her to the periphery of the community territory for days on end, sometimes months. This is dangerous because, being so close to the neighbors, they risk deadly attacks. The female may follow willingly, but oftentimes it's a forced date. It's not unusual for males to attack females, coercing them to stay close. The most telling illustration of this is the discovery of "wife-beating" equipment in one chimp community.

In Kibale Forest in Uganda, some males have taken to beating females with large wooden clubs. The first observation was an attack on a swollen female named Outamba by the highest ranking male, Imoso. Field-workers watched Imoso, who was holding a stick in his right hand, hit Outamba about five times, beating her hard. This exhausted Imoso, who took a break for a minute, after which the beating resumed. Now Imoso was handling two sticks, one in each hand, and at one point he hung from a branch above his victim, kicking her with his feet. Finally, Outamba's little daughter had had enough and came to her mother's aid, pummeling Imoso's back with her fists until he gave up.

Even though chimps have been known to use branches and sticks to hit predators, such as leopards, armed attacks on members of one's own species were, until recently, considered uniquely human. And the habit of beating females seems to have spread, because several other Kibale males have now been seen doing the same. Most attacks are directed at swollen females and always involve wooden weapons, which the investigators see as a sign of restraint. The males could also use rocks, but this might actually harm or kill their mates, which is not in their interest. They want to force obedience and usually end up mating with the females they beat.

The copycat spreading of this ugly habit shows the extent to which apes are socially influenced. They often follow the example of others. We should therefore be careful not to jump to conclusions about the "naturalness" of such behavior. Chimpanzee males are not programmed to beat females. Instead, it is something they are capable of doing under certain circumstances. Ingrained behavior is rare in our closest relatives, and it is even rarer in ourselves. There are few examples of human behavior that is both universal and develops early in life—the two best criteria for innateness. Every normal child laughs and cries, so laughing and crying seem to fit the bill. But the vast majority of human behavior does not.

Sexual coercion would, of course, be wholly unnecessary if females were ready to mate with any male. This is not the case: female chimps have outspoken partner preferences. A female may prefer to mate with a lower-ranking male despite the presence of the alpha male, who tries to control her. For days, the alpha male tries to stay close to her, not eating, having little time to drink, jealously guarding the object of his desire. As soon as he, exhausted, tries to take a nap in the middle of the day, the female will perk up and sneak off with her lover, who smartly stays within eyesight wherever she goes. I have seen alphas give up, recognizing how hopeless their enterprise is.

Male tensions can cause comical scenes. I once saw a young male named Dandy begin to make advances toward a female while restlessly looking around to see if any other males were watching. Precisely at the point when he exhibited his sexual urge by spreading his legs to show his erection to the female, one of the dominant males unexpectedly came around the corner. Dandy immediately dropped both hands over his penis, concealing it from view as if he were an embarrassed schoolboy.

And then there is the "sexual bargaining," as I have come to call it, in which chimpanzee males, instead of fighting over a female, engage in lengthy grooming sessions. One male will groom alpha for a long time before approaching a swollen female who is patiently waiting nearby. If the female is willing to mate, her partner will keep an eye on alpha while proceeding to mount her. Sometimes alpha gets up, swaying his body from side to side with his hair on end, which means trouble. The male will break off the contact with the female to do some more grooming of alpha. But after another ten minutes or so, a new attempt will be made, the male again inviting the female while keeping a watchful eye on alpha. I've also seen males get fed up with the grooming payments. Staying close to the female, they'll turn to alpha holding out their hand in the typical begging fashion of humans and apes, cupped palm up, pleading for an undisturbed mating.

Alpha himself needs to groom as well, especially when there's a tense atmosphere among the males. It's unusual for other males to band together against the top male, but the possibility can never be ruled out. The more frustrated the others become by alpha's possessiveness, the more likely it is that a mating attempt will provoke bluff displays so close to the scene that even alpha is unable to keep his mind on the sex. Grooming prices are paid by all, therefore. Strange as it may seem, male chimps groom each other most when sexual tensions run high.

## YOUNG AND NUBILE

I once took a photograph of an adolescent bonobo female grinning and squealing during copulation with a male who held two oranges, one in each hand. The female had presented herself as soon as she saw the goodies. And sure enough, she walked away from the scene with one of the two fruits. How familiar we find this pattern quickly became clear from the reaction of a professional audience to which I had shown this picture. Right after my lecture, people filed into a restaurant for lunch. A burly Australian zoologist jumped onto a table holding two oranges in the air. He got lots of laughs—our species has a quick grasp of the sexual marketplace.

A young female's self-confidence fluctuates with the size of her genital swelling. If swollen she won't hesitate to approach a male with food. She will mate with him while removing an entire bundle of branches and leaves from his hands. She will hardly give him a chance to pull out a branch for himself, sometimes grabbing the food in the midst of intercourse. This is quite a contrast to periods when she has no swelling—then she will patiently wait until he's ready to share.

Similar scenes have been witnessed in the forest. When Japanese scientists attracted bonobos to a clearing with sugarcane, adolescent females would hang around males with food, presenting their swellings. Sometimes the male would retreat, trying to avoid their advances. But the young females always persisted until they got a copulation, which invariably led to sharing. Observers noted that young females seemed to know they would get "paid" for sex. It looked as if they forced males into these transactions, since the males are not necessarily attracted to females that young.

Sex-for-food deals are known of chimps as well. Robert Yerkes, one of the pioneers of primatology, conducted experiments on what he called "conjugal" relations. After dropping a peanut between a male and a female, he would note that a swollen female's privileges exceeded those of females without such a bartering tool. Female chimps with swellings invariably claimed the prize. In nature, hunting episodes are often followed by meat-sharing with swollen females. In fact, when such females are around males may hunt more avidly because of the sexual opportunities. A low-ranking male who captures a colobus monkey automatically becomes a magnet for the opposite sex, offering him a chance to mate for meat until he's found out by someone higher-ranking than himself.

This is quite different from the quid pro quo among bonobos. Instead of the male, it is the female bonobo who seeks out these opportunities, and not just any female, only the young ones. This makes sense, given the high status of adult females, which makes trading on sex obsolete. The most intriguing part about this is that adult males give in only grudgingly to adolescent females. Do they not like the young and nubile? And if so, how does this fit what evolutionary psychologists tell us about human preferences? Male attraction to younger women is considered universal. An entire cottage industry of studies has sprung up around the theory that every man is looking for a youthful, smooth-skinned, perky-breasted, optimally fertile woman and that every woman is a gold digger, interested in men only as providers. Support comes from responses to photographs and questionnaires, whereas of course the only choices that truly matter are those made in real life and, more precisely, those leading to progeny.

Evolutionary psychologists claim that men have a precise physical standard in mind when looking for a mate. The slipper that every Cinderella needs to fit is a waistline that is 70 percent of her hip size. Known as the waist-to-hip ratio, or WHR, this 70 percent figure is thought to be programmed into human male genes. Yet this assumes an immutable male taste, whereas our species' strong point obviously is its adaptability. I find a uniform sexual taste in men as believable as the old Communist claim that one kind of car painted in one kind of color will do for the entire nation.

Beauty is in the eye of the beholder. What we find beautiful today may not always have been, which is why Peter Paul Rubens never painted a Twiggy. A recent analysis of *Playboy* playmates and Miss Americas (yes, this is what science has come to) undermines the claim of a WHR etched in stone. It shows a vast range of WHRs in modern beauty icons, from around 50 to 80 percent. If the preferred WHR has changed so dramatically over the past century, one can imagine how much more fluctuation there may have been over a longer time frame.

There is some sense, however, for a species with long-term partnerships, such as ours, to men having a bias for youthful partners. Younger women are both more available and more valuable because of the reproductive life in front of them. This bias may help explain the eternal quest of women to look young: Botox, breast implants, face-lifts, dyed hair, and so on. At the same time, we should realize how exceptional a bias this is. In bonobos and chimpanzees, males tend to go for fully mature mates. If several females are swollen at the same time, chimp males invariably hang around the older ones. They totally ignore adolescent females, even those old enough to mate and conceive. In bonobos, too, young females have to beg to get sex, whereas the older females just wait for males to come to them. Male apes practice reverse age discrimination. Perhaps they prefer mates with a track record, who already have a couple of healthy children. Within their society, such a strategy would make sense.

There is one limitation, however, that no animal can get around. To reap the advantages of reproduction, inbreeding needs to be avoided. In apes, nature's solution is female migration: young females move out, leaving all related males behind, both those she can know, such as her maternal brothers, and those she cannot know, such as her father and paternal brothers. No one assumes that apes, or other animals for that matter, have any idea of the deleterious effects of inbreeding. Migratory tendencies are a product of natural selection rather than a conscious decision: during evolutionary history, females who migrated produced healthier offspring than females who did not.

Female bonobos are not driven out by their community or abducted by neighboring males. They simply become vagabonds, hanging out increasingly at the group's periphery, breaking the tie with their mother. They enter a sexually apathic state, surely an odd state to be in for a bonobo. This way they avoid sex with community males. They leave when they are about seven, at the time that they develop their first genital swellings. Equipped with this passport, they become floaters, visiting various neighboring communities before settling down into one. Then all of a sudden their sexuality blossoms. They GG-rub with older females and copulate with new males encountered in strange forests. They now have regular, almost continuous swellings, which grow in volume with every cycle until they reach full size at the age of approximately ten. They can expect their first baby by the age of thirteen or fourteen.

For males, the situation is radically different. Investment in offspring is, in the dreary jargon of science, "asymmetrical" between the sexes. The male spends a little blob of semen, of which he has plenty. The female, in contrast, devotes one egg, which, if fertilized, results in an eight-month pregnancy requiring lots of extra food. This is followed by about five years of lactation, requiring even more extra food. If all this effort were wasted on sickly or deformed children resulting from inbreeding, the cost would be enormous. A male has far less to lose. Since a male's sisters and other females to whom he might be related are all gone or on their way out, the risk of inbreeding is minimal. Only with his own mother might this happen, and not surprisingly this is the one sexual partner combination absent in bonobo society. When her son is young, less than two years old, a mother may occasionally sexually rub against him, but soon she will stop. Having no luck with mom, juvenile males pursue sex with other females. Swollen females often accommodate the desires of these little Don Juans, who solicit them with spread legs and flicking penis. When these young males hit puberty, however, adult males begin to view them as rivals, relegating them to the fringes of foraging parties. Only many years later are they ready to claim their spot in the hierarchy. By then, their older sisters will be gone, assuring that they fertilize only unrelated females.

## VOLUPTUOUS TEMPTATIONS

Birds and fish have always held an appeal for me, so that even now my offices and labs have fish tanks, which students are sometimes asked to look after. They seek me out to learn about primates, and then I spring the fish on them! It's part of their education. Having been trained in anthropocentric disciplines, such as psychology and anthropology, they laugh at the possibility that slippery animals at the bottom of the evolutionary scale could be of interest. But they have much to teach us. And as for every creature on earth, the urge to reproduce is at the core of their existence.

In a huge tropical aquarium built into one wall of my home, one little fish made a deep impression on me. Living with many other fish, large and small, a male and female kribensis began courting. The kribensis is a pair-forming cichlid, a fish family known for its parental care. The female's belly turned purple like a ripe cherry, and both developed brilliant golden-orange fringes on their tails and dorsal fins. They quivered and danced all day and together chased off the other kribs. As usual, the male chased males, the female females. They staked out a corner of the aquarium with dense vegetation. The female's belly began to bulge. I didn't pay too much attention since fish trying to breed in a tank with a great many other fish usually lose their young to all those eager eaters around them. So I was surprised one day to find the male guarding fry.

I don't know what happened to his mate. Perhaps, in his zeal to keep his corner clean, he had kicked her out. Male care is common in cichlids, and this male was a real David against several Goliaths, keeping away fish six times his length and hundreds of times his weight. He made up for his size by pestering, butting, and certainly bothering anyone who came close. After an intruder was expelled, he would return to his collection of swimming blobs and adopt a particular body posture close to the ground that made all his children collect in a tight bunch right underneath him. With time, however, the baby fish became more venturesome, making the task of herding them harder and harder. Other fish tried to zoom in on these moving snacks, while dad worked overtime. I don't think he got anything to eat during this period, and he probably was totally stressed out. After four weeks of valiant effort, he died. From a healthy, brightly colored male, he had become a pale floater that I fished out of the tank. His offspring had grown strong enough to survive, though, and I ended up with about twenty-five kribensis, many of which I gave away.

Even though this male's life ended prematurely, it was a total success: he multiplied. From a biological standpoint, the production of offspring is worth all the work in the world. The progeny will inherit the same propensity to expend effort, resulting in a whole cycle of successful reproduction. Natural selection weeds out those who take it easy or are risk-averse: such individuals will not get many genes into the next generation. My kribensis male clearly received his genes from a long line of heroic fathers and grandfathers, faithfully carrying on their tradition.

I'm telling this fish story to make the point that deep down what we do in our societies, or what bonobos do in theirs, is not that different from what all animals do. Nowadays, of course, people limit family size—many have no children at all—but the six billion human beings in the world would not be where they are today if reproduction had not been absolutely central to our evolution. Every single human trait stems from ancestors who managed to pass on their genes. The only way our evolutionary picture differs from a fish's is that we reproduce in a far more complicated manner. We live in groups, we nurse and feed our young for years, we educate them, we seek status and privileges for them, we fight wars, we cope with inbreeding, we pass on property, and so on. Survival beyond one's own reproduction may not matter for fish, but it's an important part of our social network, explaining a curious phenomenon such as menopause as nature's way of freeing up older women to help care for their children's children. With societies far more complex than those of fish, and somewhat more complex than those of other primates, our brainpower had to expand to outsmart those around us. But we remain, at our most basic, individuals seeking the greatest possible genetic representation in the next generation.

Nature's grand theme allows us to make sense of both human and bonobo behavior and recognize that both strive for the same end by different means. In a successful bid to put a halt to infanticide, bonobos evolved a female-dominated, sexualized society in which paternity remains a mystery. When describing this society, it is hard to avoid terminology developed for our own sex lives, such as "promiscuous," "free," or "hedonic," which makes it sound as if these apes are either doing something wrong or have achieved unheard-of emancipation. Neither is the case. Bonobos simply do what they do because it provides optimal survival and reproduction in the environment in which they live.

Our own evolution took a different turn. By increasing the certainty of paternity, we paved the way for ever greater male involvement in child care. In the process, we had to limit sex outside the nuclear family: even our tiny testicles tell a story of increased commitment and curtailed freedom. Free partner switching cannot be tolerated within such a reproductive system. The taming of sexuality thus became a human obsession to the point that some cultures and religions customarily remove parts of female genitals or equate sex in general with sin. During much of Western history, the purest, most desirable human beings were the abstinent monk and the virgin nun. Suppression of the flesh is never complete, though. I find it telling that the dreams of hermits, who lived on water and stale bread, revolved around voluptuous maidens rather than good meals. For males, sex always comes first, as my chimpanzees demonstrate every time a female is swollen. They are so eager in the morning to race out of the building for a day of activity and fun that one can hold up any fruit they normally like and they will ignore it. The testosterone-filled mind is singular in its purpose.

For males, an obsession with sex may be universal, but apart from this we differ dramatically from our close relatives. We have moved sex out of the public domain and into our huts and bedrooms, to be practiced only within the family. We are by no means perfectly faithful to these restrictions, but they are a universal human ideal. The sort of societies we build and value are incompatible with a bonobo or chimpanzee lifestyle. Our societies are set up for what biologists call "cooperative breeding"—that is, multiple individuals work together on tasks that benefit the whole. Women often jointly supervise the young while men perform collective enterprises, such as hunting and group defense. The community thus accomplishes more than each individual could ever hope to accomplish on his own, such as driving a bison herd over a cliff or hauling in heavy fishnets. And such cooperation hinges on the opportunity for every male to reproduce. Each man needs to have a personal stake in the outcome of the cooperative effort, meaning a family to bring the spoils home to. This also means that men must trust each other. Their activities often remove them for days or weeks from their mates. Only if there are guarantees that nobody will get cuckolded, will men be prepared to set out together on the warpath or a hunting trip.

The dilemma of how to engender cooperation among sexual competitors was solved in a single stroke with the establishment of the nuclear family. This arrangement offered almost every male a chance at reproduction, hence incentives to contribute to the common good. We should look at the human pair-bond, therefore, as the key to the incredible level of cooperation that marks our species. The family, and the social mores surrounding it, allowed us to take male bonding to a new level, unheard of in other primates. It prepared us for large-scale collaborative enterprises that made it possible to conquer the world, from laying railroad tracks across a continent to forming armies, governments, and global corporations. In daily life we may separate the social and sexual domains, but in our species' evolution they are closely intertwined.

What makes the bonobos so appealing to us is that they have no need for any separation of these domains: they happily mix the social with the sexual. We may envy these primates for their "liberty," but our success as a species is intimately tied to the abandonment of the bonobo lifestyle and to a tighter control over sexual expressions.

FOUR

# VIOLENCE

## From War to Peace

If chimpanzees had guns and knives and knew how to handle them, they would use them as humans do.

—JANE GOODALL

I do not know with what weapons the third World War will be waged, but the fourth will be fought with sticks and stones.

—ALBERT EINSTEIN

My Georgia home offers a view of Stone Mountain, known for its huge carving of three men on horseback. The central figure, General Robert E. Lee, is so enormous that on one festive occasion long ago, forty guests had breakfast around a table placed on his granite shoulder. I have misgivings about the defenders of the South, but I have lived here long enough to question their opponents as well. Identification with the home team comes easily to group animals like ourselves. Any ill-mannered driver on the Atlanta freeway surely is one of "them Yankees."

Reminders of past violence like the Confederate Memorial Carving exist all over the world. We visit these places now with curiosity, leafing through a tourist guide, unshocked by the horror. At the Tower of London, we are told that the great philosopher Thomas More was executed and his head displayed for a month on London Bridge. In the Anne Frank House in Amsterdam, we hear about a young girl who never came home from a concentration camp. At Rome's Colosseum, we stand in the same arena where prisoners were torn apart by lions. At the Kremlin in Moscow, we admire a gilt-domed tower built by Tsar Ivan the Terrible, who enjoyed impaling and deep-frying his enemies alive. People have been killing each other forever. And we're still doing it. The security lines at airports, the bulletproof glass in taxicabs, and the emergency telephones on university campuses all tell the story of a civilization with serious problems in the live-and-let-live department.

## PLANET OF THE APES

Any civilization worth its salt has an army. We feel this so strongly that we even apply it to imaginary nonhuman civilizations, such as the one in *Planet of the Apes.* The primatologist watches the 2001 movie with horror: the cruel lead ape looks like a bipedal chimpanzee (yet sniffs like a rabbit), gorillas are depicted as dumb and obedient, an orangutan plays a slave-trader, and bonobos have conveniently been left out. Hollywood has always been more at ease with violence than with sex.

Violence reigns supreme in this movie. There is nothing as unrealistic, though, as the vast armies of uniformed apes seen on the screen. Apes lack the indoctrination, command structure, and synchronization the human military employs to intimidate its enemies. Because close coordination conveys absolute discipline, there is nothing as scary as a well-drilled army. The only other animals with armies are ants, but they lack a command structure. If army ants lose their way, such as when foragers separate from the main raiding swarm, they sometimes hook up with the tail end of their own column. They begin to follow their own pheromonal trail, forming a circular mill in which thousands of ants march and march in a densely packed circle until they all die of exhaustion. Thanks to a top-down organization, this won't happen to a human army.

Since debates about human aggressiveness invariably revolve around warfare, the command structure of armies should make us think twice before drawing parallels with animal aggression. Even if its victims understandably see military invasions as aggressive, who says that the perpetrators are in an aggressive mood? Are wars born from anger? Leaders often have economic motives, internal political reasons, or act out of self-defense. Generals follow orders, and the soldiers themselves may have no desire whatsoever to leave home and hearth. With supreme cynicism, Napoleon observed, "A soldier will fight long and hard for a bit of colored ribbon." I don't think it is an exaggeration to say that the majority of people in the majority of wars have been driven by something *other* than aggression. Human warfare is systematic and cold-blooded, making it an almost new phenomenon.

The critical word is "almost." Tendencies toward group identification, xenophobia, and lethal conflict—all of which do occur in nature—have combined with our highly developed planning capacities to "elevate" human violence to its inhuman level. The study of animal behavior may not be much help when it comes to things like genocide, but if we move away from nation-states, looking instead at human behavior in small-scale societies, the differences are not that great anymore. Like chimpanzees, people are strongly territorial and value the lives of those outside their group less than those within. It has been speculated that chimpanzees would not hesitate to use knives and guns if they had them, and similarly, preliterate people would probably not hesitate to escalate their conflicts if they had the technology.

An anthropologist once told me about two Eipo-Papuan village heads in New Guinea who were taking their first trip on a little airplane. They were not afraid to board the plane, but made a puzzling request: they wanted the side door to remain open. They were warned that it was cold up in the sky and that, since they wore nothing but their traditional penis sheaths, they would freeze. The men didn't care. They wanted to bring along some heavy rocks, which, if the pilot would be so kind as to circle over the next village, they could shove through the open door and drop onto their enemies.

In the evening, the anthropologist wrote in his diary that he had witnessed the invention of the bomb by neolithic man.

## LOATHE THY ENEMY

To know how chimps deal with strangers, one has to go to the wild. A Japanese team, led by Toshisada Nishida, has been working in Tanzania's Mahale Mountains for four decades. When Nishida invited me for a visit before his retirement, I didn't think twice. He's one of the world's greatest chimpanzee experts, and it was a treat to follow him around in the forest.

I will not go into details of life in the field camp near Lake Tanganyika—which I jokingly called the Mahale Sheraton—without power, running water, toilet, or telephone. Every day, the goal was to get up early, eat a quick breakfast, and get going as the sun rose. The chimps would need to be found, and the camp had several trackers to help out. Fortunately these apes are incredibly noisy, hence they're easy to locate. In an environment with low visibility, they rely on vocalizations. When you follow an adult male, for example, you continuously see him stop, cock his head, and listen to his group mates in the distance. Then you see him deciding how to respond, replying with his own calls, silently moving toward the source (sometimes in an incredible hurry, leaving you struggling through tangled vines), or continuing on his merry way as if what he's heard is totally irrelevant. It's well-known that chimpanzees recognize each other's voices. The forest is alive with them, some nearby, some faint in the distance, and their social life is lived largely in a world of vocalizations.

Chimps can be a boisterous, quarrelsome bunch, and on top of this they hunt. One time I was baptized while standing under a tree in which several adult males and swollen females were dividing the meat of a still-alive colobus monkey. We learned about the hunt through a sudden burst of hooting and screaming by the chimps mixed with the shrieks of monkeys. I had forgotten that during great excitement chimps often have diarrhea. Unfortunately I happened to be in the line of fire.

The next day, I saw a female walk by with a jockey-riding juvenile on her back. The daughter happily swung something fluffy in the air, which turned out to have belonged to the poor monkey. One primate's tail is another's toy. Even though chimps survive mainly on fruits and leaves, they're far more carnivorous than was once thought. They eat over thirty-five different species of vertebrates. The daily meat intake per adult chimpanzee during good times approaches that of the human hunter-gatherer during bad times. Chimps are so keen on meat, in fact, that our cook had trouble bringing a live duck from the village to camp so as to give us a break from our beans-and-rice diet. On his way, he encountered a female chimp intent on appropriating the precious bird he was holding under his arm. The brave cook withstood her threats, but just barely. Had he encountered a male chimp we would never have gotten to taste the duck.

This becomes more serious if the meal is human. Having grown up during the heyday of research, Frodo, a chimpanzee at nearby Gombe National Park, has lost all fear of people. He occasionally attacks researchers, hitting them or dragging them down a slope. But the worst incident involved a local woman, her baby, and her niece. The niece carried the woman's fourteen-month-old baby. They were crossing a small gully when they ran into Frodo, who was feeding on oil palm fronds. When the ape turned around it was too late to run. Frodo simply plucked the baby off the girl's back and disappeared. Later he was found eating the baby, who by now was dead. Baby-snatching is an extension of predatory behavior and had until then been reported outside the park only. In nearby Uganda, it's become an epidemic, with human babies being snatched from homes. Without weapons, people are helpless: wild chimpanzees can and do occasionally kill even adults of our species. Fatal attacks on humans have occurred at zoos as well.

Chimpanzees are smaller than we are. On all fours, they only reach to our knees, so people misjudge their force. You can see how muscular they are when they effortlessly scale a branchless tree. It's a feat of strength no human can replicate. The arm-pulling strength of the male chimpanzee has been measured at five times that of athletic young men, and since apes fight with four "hands," they are impossible to beat. This is so even if they're prevented from biting, as was done by a man I once met, who performed at carnivals with a chimp. Every macho guy was ready to wrestle the ape, thinking it would be a piece of cake. But even hulks the size of a pro wrestler found it impossible to control the man's partner.

One can imagine, therefore, with how much respect I gave a wide berth to chimps charging past me in the field, hair bristling, shaking small trees as they went. They were not doing this to impress me, but rather during altercations among themselves. Nothing particularly nasty happened compared to the sort of encounters that have been observed between different communities. Males regularly go on border patrols. Sometimes accompanied by females, they travel as a group toward the periphery of their territory, walking quietly in single file, alert to every sound from the other side. They may climb a tree to stare and listen for an hour or longer. Their silence seems actively enforced. If a juvenile traveling with its mother inadvertently whimpers, he or she may be threatened. Everyone in the patrol is on edge. A snapping twig or the sudden sound of a running bushpig causes them to grin nervously and seek reassurance from each other, touching and embracing. They relax only upon return to safer parts of their territory, releasing tension in bursts of calling and drumming.

Given how chimpanzees of different communities treat each other, I would be nervous, too. Males kill each other through highly coordinated actions against single individuals of another community—they will stalk, run down, and swiftly overwhelm a victim, who is beaten and bitten so viciously that he either dies on the spot or has no chance of survival. A few such surprise attacks have been actually witnessed, but most of the time the evidence consists of gruesome finds in the forest. At some research sites, no dead bodies have been found, but healthy males have one by one vanished from a community until none remained.

At the Mahale Mountains, Nishida witnessed border patrols and violent charges against strangers. He believes that all the males of one of his communities were gradually wiped out by neighboring males over a period of twelve years. The winners then claimed the vacant territory and the females in it. There is no question that chimpanzees are xenophobic. In one attempt to introduce ex-captive chimps back into the forest, the local wild chimps reacted so viciously that the project had to be abandoned.

Given their enormous territories, violent incidents between chimp communities are rarely witnessed. The few instances that have been seen leave little doubt, though, that we're dealing with targeted, deliberate killing—in other words, "murder." Realizing how controversial such a claim would sound, Jane Goodall asked herself where the impression of intentionality came from. Why couldn't the killing be a mere side effect of aggression? Her answer was that the attackers showed a degree of coordination and abuse not seen during aggression within their own community. The chimps acted almost the way they do toward prey, treating the enemy as if it belonged to another species. One attacker might pin down the victim (sitting on his head, holding his legs) while others bit, hit, and pounded. They would twist off a limb, rip out a trachea, remove fingernails, literally drink blood pouring from wounds, and in general not let up until their victim stopped moving. There are reports of attackers returning to the scene weeks later, seemingly to check on the outcome of their efforts.

This frightening behavior, sadly, is not unlike that of our own species. We routinely *dehumanize* our enemies—like the chimps, treating them as less than one of our own species. During the first weeks of the Iraq war, I was struck by an interview with an American pilot, who enthusiastically explained that as a young boy he'd followed the Gulf War and had been fascinated by the precision bombing. He could not believe that he himself was now using even more sophisticated smart bombs. For him, the war was all about technology; it was like a computer game that he was finally permitted to play. What happened at the other end didn't seem to even enter his mind, which is perhaps just what the military wants. Because as soon as one begins to see the enemy as human, things begin to fall apart.

Us-versus-them thinking comes remarkably easily to us. In one psychological experiment, people were randomly assigned different colored badges, pens, and notepads and simply labeled the "Blues" and the "Greens." All they were asked to do was evaluate each other's presentations. They liked the presentations by people with their own color designation the best. In a more elaborate fabrication of group identity, students were assigned the roles of guards and prisoners in a prison game. They were supposed to spend two weeks together in a basement at Stanford University. Six days into it, however, the experiment had to be broken off because the "guards" had become increasingly arrogant, abusive, and cruel and the "prisoners" started to revolt. Had the students forgotten that it was just an experiment and that their roles had been decided by the flip of a coin?

The Stanford Prison Experiment gained notoriety when it came to light that American officers had tortured detainees in Baghdad's Abu Ghraib prison. American guards employed a wide range of torture techniques, including hooding and attaching electric wires to prisoners' genitals. Some in the U.S. media tried to downplay these events as "pranks," but dozens of prisoners actually died in the process. Apart from the striking similarities with the brutality and sexual undertones in the Stanford Prison Experiment, the guards and the prisoners at Abu Ghraib were of different races, different religions, and spoke different languages. This made their dehumanization even easier for the guards. Janis Karpinski, the general responsible for the military police there, said she'd been ordered to treat prisoners "like dogs." One of the horrific pictures to come out of this prison indeed showed a female officer dragging a naked prisoner across the floor with a leash around his neck.

The in-group always finds reasons to see itself as superior. The most extreme historic example of this tendency is, of course, Adolf Hitler's creation of an out-group. Depicted as less than human, the out-group enhances the solidarity and self-worth of the in-group. It's a trick as old as man, but the psychology may in fact precede our species. Beyond just the identification with a group, which is widespread in animals, are two other characteristics we have in common with chimps. The first, as we have seen, is a loathing of the out-group to the point of dehumanization (or "dechimpization"). The gulf between in-group and out-group is so huge that aggression falls into two categories: one contained and ritualized within the group, the other all-out, gratuitous, and lethal between groups.

The other even more disturbing out-group phenomenon that emerged at Gombe involved chimps who actually knew each other. Over the years, one community split into a northern and southern faction, eventually becoming separate communities. These chimpanzees had played and groomed together, reconciled after squabbles, shared meat, and lived in harmony. But the factions began to fight nonetheless. Shocked researchers watched as former friends now drank each other's blood. Not even the oldest community members were left alone. An extremely frail-looking male, Goliath, was pummeled for twenty minutes and dragged about. Any association with the enemy was grounds for attack. If patrolling chimpanzees found fresh sleeping nests in a tree in the border region, they displayed around, pulling the branches apart until all enemy nests were destroyed.

So us-versus-them among chimpanzees is a socially constructed distinction in which even well-known individuals can become enemies if they happen to hang out with the wrong crowd or live in the wrong area. In humans, ethnic groups that used to get along reasonably well may all of a sudden turn against each other, as the Hutus and Tutsis did in Rwanda and the Serbs, Croats, and Muslims in Bosnia. What kind of mental switch is flipped that changes people's attitudes? And what kind of switch turns chimpanzee group mates into each other's deadliest foes? I suspect the switches operate similarly in humans and apes and are controlled by the perception of shared versus competing interests. So long as individuals feel a common purpose, they suppress negative feelings. But as soon as the common purpose is gone, tensions rise to the surface.

Both humans and chimps are gentle, or at least restrained, toward members of their own group, yet both can be monsters to those on the outside. I'm simplifying, of course, because chimps can also kill within their own community, as can people. But the in-group versus out-group distinction is fundamental when it comes to love and hate. This is true also for captive apes. At the Arnhem Zoo, the chimpanzees developed a habit of patrolling even though there were no enemy groups. In the late afternoons, a few males would begin walking around the large island, eventually followed by all the adult males and some juveniles. Obviously, they didn't show any of the tensions seen in patrols in the wild, but this behavior indicates that territorial borders have a meaning even under artificial circumstances.

Captive chimps are just as xenophobic as those in the wild. It's almost impossible to introduce new females to an existing zoo group, and new males can be brought in only after residential males have been removed. Otherwise, a bloodbath results. The last time we tried male-switching at the Yerkes Primate Center, the females kicked out the first few new males (meaning the females attacked them, and we had to withdraw them to save their lives).

Several months later, we tried two new males. One of them received the same unpleasant reception as before, but the other, named Jimoh, was permitted to stay. Within minutes of Jimoh's introduction, two older females contacted him and groomed him, after which they fiercely defended him against every other female. Years later, during a background check on our chimps, I discovered that Jimoh had not been as new as we thought. Fourteen years before his introduction to our group, Jimoh had lived at a different institution with the same two females who now protected him. They hadn't seen each other in the meantime, but this earlier contact from long ago had made all the difference.

## MINGLING AT THE BORDER

Does the fact that one of our closest relatives kills its neighbors mean that, as a recent documentary put it, "warfare is in our DNA"? This makes it sound as if we are destined to be a warring people forever. But even ants, which definitely have warlike DNA, are not violent as long as they have plenty of space and food. What would be the point? It's only when one colony's interests collide with those of another that such behavior makes sense. War is not an insuppressible urge. It is an option.

Nevertheless, it cannot be coincidental that the only animals in which gangs of males expand their territory by deliberately exterminating neighboring males happen to be humans and chimpanzees. What is the chance of such tendencies evolving independently in two closely related mammals? The human pattern most similar to that of the apes is known as "lethal raiding." Raids consist of a group of men launching a surprise attack when they have the upper hand, hence when there's little chance that they will suffer themselves. The goal is to kill other men and abduct women and girls. Like the territorial violence among chimpanzees, human raids are not exactly acts of bravery. Surprise, trickery, ambush, and the avoidance of daylight are favored tactics. The majority of hunter-gatherer societies follow this pattern, waging war every couple of years.

But does the prevalence of lethal raiding imply, as stated by Richard Wrangham, that "chimpanzee-like violence preceded and paved the way for human war, making modern humans the dazed survivors of a continuous, 5-million-year habit of lethal aggression"? The problem word here is not "dazed," which is mere hyperbole, but "continuous." For this to be true, our earliest ancestor would need to have been chimpanzeelike and we must have been on the warpath ever since. There is no evidence for either assumption. First, since the split between humans and apes, apes have undergone their own evolution. No one knows what happened during those five to six million years. Due to poor fossilization in forests, our record of ape ancestry is sketchy. The last common ancestor of humans and apes may have been gorillalike, chimpanzeelike, bonobolike, or different from any living species. Not too different, of course, but we certainly have no proof that this ancestor was a warmongering chimpanzee. And it's good to keep in mind that only a handful of chimp populations have been studied, and not all of them are equally aggressive.

Second, who says that our ancestors were as brutal as we are today? Archeological signs of warfare (protective walls around dwellings, graveyards with weapons embedded in skeletons, depictions of warriors) go back only ten to fifteen thousand years. In the eyes of the evolutionary biologist, this is recent history. On the other hand, it's hard to believe that war appeared out of nowhere without previous hostilities between human groups. Some proclivity must have existed. Most likely, territorial aggression was always a potential, but one exercised on a small scale only, perhaps until man settled down and began to accumulate possessions. This would mean that, instead of having waged war for millions of years, we first knew sporadic intergroup conflict, which only recently developed into actual warfare.

But it's hardly surprising that scientists emphasizing man's violent side have flocked to the chimpanzee as Exhibit A. The parallels are undeniable and disturbing. Though one aspect of human behavior that the chimp cannot illuminate is something we do even more than wage war: maintain peace. Peace is common among human societies, as is the trading of goods, the sharing of river water, and intermarriage. Here chimps have nothing to tell us, since they lack any friendly ties between groups. All they know is varying degrees of hostility. This means that to understand human intergroup relations at a primal level, we need to look beyond the chimpanzee as an ancestral model.

There's an intriguing remark by famous entomologists Bert Hölldobler and Ed Wilson, in *Journey to the Ants*, about the existence of two kinds of scientists. The theoretician is interested in a particular issue and looks for the best organism to resolve it. Geneticists have picked the fruit fly and psychologists the rat. They're not really interested in fruit flies or rats, only in the problems they wish to solve. The naturalist, on the other hand, is interested in a particular class of animals for their own sake, realizing that each animal has its own story to tell, one that will prove of theoretical interest if one studies hard enough. Hölldobler and Wilson count themselves in the latter category, as do I. Instead of focusing on human aggression as *the* issue, hence the chimpanzee as *the* species, as has been done ever since the killer ape theory came along, my attention is drawn to a less brutal ape on the sidelines of this debate. And the behavior of this ape illuminates a different capacity: the capacity for peace.

Peaceful mingling between bonobo groups was first noted in the 1980s when different communities came together in Wamba Forest in the Democratic Republic of the Congo and stayed together for an entire week before splitting up again. This may hardly seem spectacular, but the event was as shocking as the violence among former chimpanzee friends in Gombe. It countered the persistent belief that our lineage is naturally violent. I once saw a video of intergroup mingling in which bonobos fiercely chased each other at first, screaming and barking, but without any physical contact. Then, gradually, females of different groups engaged in GG-rubbing and even groomed one another. In the meantime, their offspring played and wrestled with age peers. Even males of opposing camps eventually engaged in brief scrotal rubbing.

In more than thirty separate intergroup encounters at Wamba, members of the opposite sex typically met in a sexual and friendly manner. Males, on the other hand, were generally hostile and standoffish toward males of the other group. Copulations between males and females of different groups were common during the first fifteen minutes of an encounter.

At a different bonobo site, in Lomako Forest, similar observations were made. Males of different groups sometimes wildly chased each other through the undergrowth while the females hung in the trees, shouting and screaming. The clashes looked so fierce that field-workers who were watching got goose bumps. But afterward, the bonobos would be unscathed, and intergroup mergers would occur. They would start out tense, but then the apes would settle down and engage in sex and grooming between the two communities. Only the males of different groups failed to have friendly contact.

There also were days on which bonobos did not wish to mingle with their neighbors and kept them at a distance. Field-workers might be startled by sudden drumming sounds and bonobos dropping out of the trees. Then the apes would rush toward the members of another group, shouting and carrying on. At the border of their respective territories, the groups would sit in the trees calling at each other. It should be stressed, though, that despite occasional injuries after skirmishes between communities, not a single fatality has been recorded.

The overlapping ranges and mingling at the borders of bonobo communities stand in stark contrast to how chimp groups interact. When the mist lifts from the evolutionary pressures that shaped bonobo society, perhaps we will understand how they have managed to escape what many people consider the worst scourge of humanity: our xenophobia and our tendency to discount the lives of our enemies. Is it because bonobos fight, if they fight at all, not for a fatherland, but for a motherland? Males of any species naturally try to monopolize females, but once female bonobos achieved the upper hand, males may have lost control to the extent that females copulate freely with whomever they want, including neighbors. This made male territorial competition obsolete. First, sexual mingling of course translates into reproduction, which means that neighboring groups may include your relatives: enemy males may be brothers, fathers, and sons. And second, it makes no sense for males to risk life and limb to get to females who are already happy to have sex with them.

Bonobos show us the conditions under which peaceful relations between groups may evolve. Similar conditions apply to us. All human societies know intermarriage, hence gene flow between groups, which makes deadly aggression counterproductive. Even though one may gain by defeating another group over territory, there are drawbacks, such as lives lost on your own side, kin killed on the other side, and reduced trading opportunities. The latter may not apply to apes, but is a significant factor in the human case. Our intergroup relations are therefore inherently ambivalent: a hostile undertone is often combined with a desire for harmony. The bonobo nicely illustrates the same ambivalence. Their neighborhood relations are far from idyllic—they seem to take every opportunity to underline territorial boundaries—yet they keep the door open to de-escalation and friendly contact.

Even if female migration creates genetic exchange in chimpanzees, hostility between their communities precludes the free sexual relations seen in bonobos. No one knows what came first—the absence of reproduction between groups or the severe hostility—but the two obviously amplify each other, thus creating a perpetual cycle of violence among chimps.

The upshot is that humans share intergroup behavior with both chimps and bonobos. When relations between human societies are bad, they are worse than between chimps, but when they are good, they are better than between bonobos. Our warfare exceeds the chimpanzee's "animal" violence in alarming ways. But at the same time the payoffs from neighborly relations are richer than in bonobos. Human groups do a lot more than mingle and have sex. They exchange goods and services, have ceremonial feasts, allow one another to travel through, and arrange common defenses against hostile third parties. When it comes to intergroup relations, we beat our close relatives on both the positive and the negative end of the scale.

## GIVE PEACE A CHANCE

Upon my arrival from Europe more than two decades ago, I was taken aback by the amount of violence in the American media. I don't just mean the daily news, but everything from sitcoms, comedies, and drama series to movies. Staying away from Schwarzenegger and Stallone hardly does it: almost any American movie features violence. Inevitably, desensitization sets in. If you say, for example, that *Dances with Wolves* (the 1990 movie with Kevin Costner) is violent, people look at you as if you are crazy. They remember an idyllic, sentimental movie, with beautiful landscapes, about a rare white man who respected American Indians. The blood and gore barely register.

Comedy is no different. I love *Saturday Night Live* for its inside commentary on peculiarly American phenomena such as cheerleaders, televangelists, and celebrity lawyers. But *Saturday Night Live* is incomplete without at least one sketch in which someone's car explodes or head gets blown off. Characters like Hans and Franz appeal to me for their names alone (and yes, I do have a brother named Hans), but when their free weights are so heavy that their arms get torn off, I'm baffled. The spouting blood gets a big laugh from the audience, but I fail to see the humor.

Did I grow up in a land of sissies? Perhaps, but the important point is that there's a big difference in how violence is portrayed in different societies. And which do we value most: harmony or competitiveness? This is the problem with the human species. Somewhere in all of this resides a true human nature, but it's stretched in so many different directions that it's difficult to say whether we're naturally competitive or naturally community-building. In fact, we are both, but each society reaches its own balance. In

America, "the squeaky wheel gets the grease." In Japan, "the nail that stands out gets pounded into the ground."

Does this variability mean that we cannot learn from other primates? It's not that simple. First, each species has its own way of handling conflict. Chimps are more confrontational than bonobos. But also within each species, like in humans, we find variation from group to group. We see "cultures" of violence and "cultures" of peace. And the latter are made possible by our universal primate ability to iron out differences.

I'll never forget one particular winter day at the Arnhem Zoo. The entire chimpanzee colony was locked indoors, out of the cold. In the course of a charging display, I watched the alpha male attack a female, which caused great commotion as other apes came to her defense. The group calmed down, but an unusual silence followed, as if everyone was waiting for something. It lasted a couple of minutes. Then, unexpectedly, the entire colony burst out hooting, and one male rhythmically stamped on the metal drums stacked in the corner of the hall. In the midst of all the pandemonium, at the center of attention, two chimps kissed and embraced.

I reflected on this sequence for hours before I realized that the two embracing apes had been the male and female from the original row. I know, I'm slow, but no one before had ever mentioned the possibility of *reconciliation* in animals. At least, this was the term that came to my mind. Since that day, I've been studying peacemaking, or as we call it nowadays, conflict resolution, among chimps and other primates. Others have done the same in a variety of species, including dolphins and hyenas. It seems that many social animals know how to reconcile, and for good reason. Conflict is inevitable, yet at the same time animals depend on one another. They forage for food together, warn each other of predators, and stand united against enemies. They need to maintain good relationships despite occasional flare-ups, just like any married couple.

Golden monkeys do it with hand-holding, chimpanzees with a kiss on the mouth, bonobos with sex, and tonkean macaques with clasping and lipsmacking. Each species follows its own peacemaking protocol. Take something I've seen repeatedly during reconciliations among apes but never monkeys: after one individual has attacked and bitten another, he or she returns to inspect the inflicted injury. The aggressor knows exactly where to look. If the bite was aimed at the left foot, the aggressor will without hesitation reach for the victim's left foot—not the right foot or an arm—lift and inspect it, then begin cleaning the injury. This suggests an understanding of cause and effect along the lines of "If I have bitten you, you must now have a gash in the same spot." It suggests that the ape takes another's perspective, realizing the impact of its own behavior on somebody else. We may even speculate that they regret their actions, just as we often do. The German naturalist Bernhard Grzimek experienced this after having had the luck to survive a vicious attack by a male chimpanzee. When his rage had died down, the ape seemed extremely concerned about Grzimek. He approached the professor and tried, with his fingers, to close and press together the edges of the worst wounds. The undaunted professor let him do so.

The definition of reconciliation (a friendly reunion between opponents not long after a fight) is straightforward, but the emotions involved are hard to pinpoint. The least that occurs, but this is already truly remarkable, is that negative emotions, such as aggression and fear, are overcome in order to move to a positive interaction, such as a kiss. The bad feelings are reduced or left behind. We experience this transition from hostility to normalization as "forgiveness." Forgiveness is sometimes touted as uniquely human, even uniquely Christian, but it may be a natural tendency for cooperative animals.

Only animals without memories could possibly ignore conflict. As soon as social events are stored in long-term memory, as in most animals and humans, there is a need to overcome the past for the sake of the future. Primates form friendships, expressed in grooming, traveling together, and defending each other. That fights create anxiety about the state of the relationship is suggested by an unexpected indicator. Just as college students scratch their heads during a tough exam, self-scratching in other primates indicates unease. If one takes notes on self-scratching, as some researchers have done, it turns out that both parties involved in a fight scratch themselves a lot, but stop after having been groomed by their opponent. We can surmise that they were worried about their relationship and reassured by the reunion.

People raising young apes at home say that, after a reprimand for unruly behavior (the only behavior young apes seem to know), there is an overwhelming desire to patch things up. The ape sulks and whimpers until he can't stand it anymore. Then he jumps into his adoptive parent's lap, wrapping both arms around her, squeezing the air out of her. This is often followed by an audible sigh of relief once he's being comforted by the parent.

Primates learn peacemaking early in life. As with everything related to attachment, it starts with the mother-infant bond. During weaning, the mother pushes the infant away from her nipples, yet allows it to return right away when it screams in protest. The interval between rejection and acceptance lengthens with the infant's age, and conflicts turn into major scenes. Mother and offspring bring different weapons to this battlefield. The mother has superior strength and the offspring a well-developed larynx (a juvenile chimpanzee easily out-screams several human children) and equally well-developed blackmailing tactics. The youngster will cajole mother with signs of distress, such as pouts and whimpers, and if all else fails, a temper tantrum at the peak of which he may almost choke in his screams or vomit at her feet. This is the ultimate threat: a literal waste of maternal investment. One wild mother's answer to these histrionics was to climb high up into a tree and throw her son to the ground, or so it seemed, while at the last instant holding on to his ankle. The young male hung upside down for fifteen seconds, screaming his head off, before his mother retrieved him. There were no more tantrums that day.

I've seen fascinating compromises, such as in one youngster who sucked on his mother's lower lip. The lip-locked male, already five years old, had settled for this substitute. Another juvenile stuck her head under her mother's arm, quite close to the nipple, to suck on a skin-fold. These compromises last only a few months, after which the youngster moves on to solid food. Weaning conflict is life's first negotiation conducted with a social partner absolutely needed for survival. It contains all the right ingredients: conflicting interests, overlapping interests, and a cycling through positive and negative encounters that results in some sort of compromise. Maintaining the all-important tie with mother despite discord lays the groundwork for later conflict resolution.

Reconciliations with peers are next in importance and are learned early in life as well. While watching a large outdoor group of rhesus monkeys, I observed the following little scene. Oatly and Napkin, two four-month-olds, were play-wrestling when they were joined by Napkin's adult aunt. The aunt "helped" Napkin by holding her playmate down. Napkin took advantage of the unequal situation by suddenly jumping on Oatly and biting her. After a brief struggle, they broke up. The incident was not too serious, but its aftermath was remarkable. Oatly walked straight up to Napkin, who was sitting with the same aunt, and groomed her back. Napkin turned around, and the two infants embraced belly-to-belly. To complete this cozy picture, the aunt then put both arms around them.

This happy ending caught my attention not just because the infants were still so young and tiny (comparable to human toddlers) but also because rhesus monkeys are probably the worst reconcilers. They're nasty characters with strict hierarchies in which the higher-ups rarely hesitate to punish those lower on the scale. The species will not be nominated for the Primate Peace Prize anytime soon. But perhaps there is hope given the outcome of a crazy idea I got after a lecture to a roomful of child psychologists. I had taken my audience to task about the fact that we know more about reconciliation in other primates than in our own species. This remains true today. Psychologists tend to focus on abnormal or problematic behavior, such as bullying, so that we know startlingly little about the spontaneous, normal ways in which conflict is reduced or overcome. In defense of this lamentable situation, one scientist in the room argued that human reconciliation is far more complex than in monkeys, influenced as it is by education and culture. In other primates, he said, it's mere instinct.

But the word "instinct" stuck in my mind. I barely know what this means anymore, since purely inborn behavior is impossible to find. Like humans, other primates develop slowly; they have years to be influenced by the environment in which they grow up, including its social fabric. In fact, we know that primates adopt all sorts of behaviors and skills from each other, and therefore groups of the same species may act quite differently. No wonder primatologists increasingly speak of "cultural" variability. Most of this variability concerns tool-use and eating habits, such as chimpanzees cracking nuts with stones or Japanese monkeys washing potatoes in the ocean. But *social* culture is a distinct possibility as well.

This discussion with the psychologists gave me an idea. I put juveniles of two different macaque species together for five months. The typically quarrelsome rhesus monkeys were housed with the far more tolerant and easygoing stumptail monkeys. After a fight, stumptail monkeys often reconcile by holding each others' hips in a so-called hold-bottom ritual. Surprisingly, the rhesus monkeys were afraid at first. Not only are stumptails slightly larger, the rhesus must have sensed a toughness underneath their gentle temperament. So, with the rhesus clinging in a fearful cluster to the ceiling of the room, the stumptails calmly inspected their new environment. After a couple of minutes, a few rhesus, still in the same uncomfortable position, dared threaten the stumptails with some harsh grunts. If this was a test, they were in for a surprise. Whereas a dominant rhesus monkey would have answered the challenge in no uncertain terms, the stumptails simply ignored it. They didn't even look up. For the rhesus monkeys, this must have been their first experience with dominant companions who felt no need to assert their position.

During the study, the rhesus learned this lesson a thousand times over and also engaged in frequent reconciliations with their gentle oppressors.

Physical aggression was highly exceptional and the atmosphere was relaxed. By the end of the five months, the juveniles played together, groomed together, and slept in large mixed huddles. Most important, the rhesus monkeys developed peacemaking skills on a par with those of their more tolerant group mates. At the end of the experiment, after we separated the species, the rhesus monkeys continued to show three times more friendly reunions and grooming after fights than was typical of their kind. Jokingly, we called them our "New and Improved" rhesus monkeys.

This experiment showed that peacemaking is an acquired social skill rather than an instinct. It's part of social culture. Each group reaches its own balance between competition and cooperation. This is as true for monkeys as it is for people. I'm from a culture that is characterized by consensus-building, perhaps because the Dutch live at high density on land wrested from a formidable common enemy, the North Sea. Other countries, such as the United States, encourage individualism and self-reliance rather than group loyalty. This may have to do with mobility and empty space. In the old days, if people didn't get along, they could always settle elsewhere. Conflict resolution may not have been emphasized to the degree preferable now that the United States has become a more crowded place. Science should study the skills that normally prevent the escalation of conflict and keep aggression in check. Do we teach our children to stand up for themselves or to find mutually agreeable solutions? Do we teach them rights or responsibilities? Human cultures show incredible contrasts in this regard, and a recent discovery shows similar variability among wild primates.

Like rhesus monkeys, olive baboons have a fierce reputation. They're not the sort of primates one would expect to go the flower-power route, but this is exactly what happened with one troop in the Masai Mara in Kenya. Every day, males of a troop studied by American primatologist Robert Sapolsky fought their way through the territory of another troop to get access to the garbage pit of a nearby tourist lodge. Only the biggest and meanest males would make it through. The bounty was definitely worth fighting over until the day the lodge discarded meat infected with bovine tuberculosis. It killed off all the baboons that ate it. This meant that the troop under study lost many males, and not just any males, but the most aggressive ones. As a result, the troop suddenly became an unlikely oasis of harmony and peace in the harsh world of baboons.

This by itself was hardly surprising. The number of violent incidents in the troop naturally dropped after the bullies got wiped out. It became more interesting when it was discovered that this pattern was maintained for a decade, even though by then *none* of the troop's original males were around anymore. Baboon males migrate after puberty, hence fresh young males enter troops all the time. So, despite a complete turnover of males, this particular troop upheld its pacifism, tolerance, increased grooming, and exceptionally low stress levels. How the tradition had been maintained remains unclear. Female baboons stay all their lives in the same troop, so their behavior probably holds the key. Perhaps they had become selective in their acceptance of new males or managed to perpetuate the relaxed atmosphere of the early years by grooming more with males, relaxing them. We don't have the answer, but two chief conclusions from this natural experiment are loud and clear: behavior observed in nature may be a product of culture, and even the fiercest primates do not forever need to stay this way.

Perhaps this applies to us as well.

## GIRLS' WHISPERINGS

"Who can one hit, if not one's friends?" one British wit exclaimed to another before punching him in the jaw.

Those Brits are funny, but it's actually not unusual for men to mix friendship and rivalry. The gulf between the two is not nearly as wide for men as for women—at least this is my opinion after a lifetime of "studying" people as a participant observer. Unfortunately, the way people resolve conflicts is barely a topic of research. Are women better at it? Are men by definition warriors? Men and women have been assigned to different planets, Mars and Venus, but is it that simple? Everywhere in the world, the murder rate for men far exceeds that for women, and wars are typically fought by men, so it seems fair to blame the Y chromosome for the mess we're in. If women have an edge when it comes to pacifism, however, it may not be because they're good at repairing what has been broken. I see the strength of women in conflict prevention and their distaste for violence. But they're not necessarily good at the diffusion of tensions once these have arisen. The latter is, in fact, a male forte.

Chimpanzee females have far fewer fights than males, probably because they work hard to avoid them. If a fight does occur, however, females rarely reconcile. At the Arnhem Zoo, males reconciled nearly half their confrontations, females only one in five. A similar difference has been observed in the field. Males cycle through fights and reunions, whereas females have a preemptive attitude toward conflict. Unlike males, they take care to stay on good terms with those with whom they enjoy close ties, such as offspring and best friends, and they let aggression run its ugly course when it comes to their rivals. On a recent visit to Arnhem, I found Mama and Kuif grooming as if time stood still: they were already friends three decades ago. I remember times when Mama favored one political "candidate" among the males and Kuif another, and I marveled at the way they acted as if they didn't notice the other's choice. Mama would make a wide detour during power struggles among the males to avoid coming face-to-face with her friend, who had joined the rival's ranks. Given Mama's undisputed dominance and extreme temper toward females who failed to obey her, her lenience toward Kuif was an amazing exception.

But on the negative side, females can be incredibly mean and calculating. A good example is deceptive reconciliation offers. The idea here is to trap the opponent using false pretenses. Puist, a heavy, older female, pursues and almost catches a younger opponent. After her narrow escape, the victim screams for a while, then sits down, panting heavily. The incident seems forgotten, and ten minutes later Puist makes a friendly gesture from a distance, stretching out an open hand. The young female hesitates at first, then approaches Puist with classic signs of mistrust, such as frequent stopping, looking around at others, and a nervous grin on her face. Puist persists, adding soft pants when the younger female comes closer. Soft pants have a particularly friendly meaning; they are often followed by a kiss, the chimpanzee's chief conciliatory gesture. Then, suddenly, Puist lunges and grabs the younger female, biting her fiercely before she manages to free herself.

Reconciliations among male chimps may be edgy, sometimes even unsuccessful (meaning that the fight starts all over again), but they never include trickery. Males carry their tensions on their sleeves. Between close buddies, such as Yeroen and Nikkie during their joint rule, one male may get upset if his friend does something he dislikes, such as inviting a sexually attractive female. He puts his hair up and starts swaying his upper body, hooting softly, giving off an immediate signal that something is amiss. If the other objects to these objections, a confrontation will erupt, which most of the time will be quickly reconciled. Male chimps make up easily in contrast to females, among whom tensions tend to linger. It's not unusual for two female chimps to meet and suddenly start screaming and yelling at each other, even though I, the observer, have not the slightest clue as to what may have triggered the outburst. These incidents give the impression that something has been brewing under the surface, perhaps for days or weeks, and that I just happen to be present when the volcano erupts. This never occurs among males, mainly because males signal hostilities and disagreements openly so that things are always "talked" out, one way or another. This may involve overt aggression, but at least the air is cleared.

Bonobo females reconcile far more readily than their chimpanzee counterparts. Asserting collective dominance and relying on an extensive network of alliances induces a need for female solidarity. Without carefully servicing their ties, they could never stay on top. Conversely, bonobo males reconcile less than chimpanzee males. Here again, the reason is a practical one: bonobo males lack the intense cooperation in hunting, political alliances, and territorial defense that forces male chimpanzees to preserve unity. The tendency to reconcile, then, is a political calculation that varies by species, by gender, and by society. Paradoxically, the level of aggression says little about peacemaking: the more aggressive gender may be best at making peace and the more peaceful gender worst. The popular Mars versus Venus distinction makes it seem as if there's only one dimension to consider, but both apes and people are far too complex for that.

The main reason for peacemaking is not peace per se, but shared purpose. This can be seen after common trauma. For example, after the 9/11 attack on the World Trade Center in New York, tensions between the races in the city dropped. Nine months after September 2001, when asked how they saw race relations, New Yorkers of all races called these relations more often good than bad. In the years before, they had overwhelmingly called them more often bad than good. The post-attack feeling of "we're in this together" fostered exceptional unity and made people far more accepting and conciliatory than usual. Ethnic out-groups were all of a sudden seen as belonging to a citywide in-group.

This makes sense in view of theories about why reconciliation evolved in species as diverse as hyenas, baboons, and humans. Mutual dependency fosters harmony. There was a time when biologists cared only about winning and losing: winning was good, losing bad. Every population had its "hawks" and "doves," and the doves had a tough time staying alive. The problem is that who wins and who loses is only half the story. If one's livelihood depends on working together, as it does for myriad animals, those who initiate fights risk losing something far more important than the conflict at hand. Sometimes one cannot win a fight without losing a friend. In order to be successful, social animals need to be hawks *as well as* doves. New theories emphasize reconciliation, compromise, and the need for good relationships. In other words, patching things up is not done for the sake of being nice, but in order to maintain cooperation.

In one study, monkeys were trained to work together. They could eat from a popcorn machine provided they arrived in pairs. Coming alone would get them no snacks at all. They had no trouble learning this. After this training, fights were induced to see how quickly these monkeys would reconcile. Pairs of monkeys that had come to depend on one another increased their reconciliation rate manifold. The mutually dependent monkeys clearly grasped the advantage of staying on good terms.

Obviously, this principle is familiar to people, too. In fact, it is the ideal underlying the European Union, which grew out of the European Community founded in the 1960s. After centuries of warfare on the continent, some visionary politicians argued that fostering economic ties between nations might be the solution: there would be too much at stake to continue the same behavior. Like the monkeys trained to feed together, Europe's national economies now feed off each other. By invading another country, a nation would only hurt its own economy. This disincentive to warfare has held for over half a century.

Pragmatic solutions to conflict, such as formation of the European Union, are typically male. I say this without chauvinism, equally aware that males are also responsible for the worst excesses of violence when peace attempts fail. One of the very few studies on the different ways genders manage disagreements focused on children's games. It found girls playing in smaller groups and less competitively than boys. The average girls' game didn't last long, however, because girls were not nearly as good as boys at resolving disputes. Boys quarreled all the time, debating the rules like little lawyers, but this never meant the end of the game. After an interruption, they would simply continue. Among girls, however, a quarrel usually meant that the game broke up. No efforts would be made to get the team back together again.

The nature of girls' and boys' disputes is different as well. Let us say individual A walks up to B, whereupon B turns away and acts as if A does not exist. Could you imagine boys seeing this as a fight? They'd simply move on to something else. For two girls, on the other hand, such an encounter may be excruciating. It might reverberate for hours or days. Counting fights in the school yard, a Finnish research team found far fewer episodes among girls than boys. This was as expected, but when they asked each child by the end of the day if he or she had been in a fight, they suddenly had roughly equal numbers of fights for boys and girls. Aggression among girls is often barely visible to the naked eye. In her novel *Cat's Eye,* Margaret Atwood contrasted the torments to which girls subject one another with the straightforward competition among boys. Her principal character complains:

> *I considered telling my brother, asking him for help. But tell what exactly? Cordelia does nothing physical. If it was boys, chasing or teasing, he would know what to do, but I don't suffer from boys in this way. Against girls and their indirectness, their whisperings, he would be helpless.*

This kind of subtle aggression does not fade easily, as reported by the Finnish researchers. They found that discord among girls outlasted that among boys. If asked how long they might stay angry with each other, boys thought in terms of hours, sometimes days, whereas girls felt they might stay angry for the rest of their lives! Grudge-holding makes for erosive relationships, as explained by a swimming coach interviewed about her switch from training women to training men. The coach found working with the opposite sex far less stressful. She explained that if two young women got into a fight at the beginning of the season, there'd be little chance of their getting over it for the remainder of the year. The fight would be festering day after day, sapping the team's solidarity. Young men, on the other hand, had tons of fights. But in the evening they would drink a beer together, and the next day they would barely remember the fight.

For boys and men, rivalry and hostilities do not stand in the way of good relations. In *You Just Don't Understand,* linguist Deborah Tannen reports hostile conversations followed by friendly chats among men. Men use conflict to negotiate status and actually enjoy sparring, even with friends. When things get heated, men often find a way of making up afterward with a joke or apology, thus maintaining bonds by alternating between camaraderie and mild hostility. For example, businessmen may shout and bully at a meeting, only to take a restroom break during which they joke and laugh. "Nothing personal," is a typical male remark after a harsh exchange.

If conflict is like bad weather, women try to stay out of it, whereas men buy an umbrella. Women are peacekeepers, men peacemakers. Women's friendships are often seen as more profound and intimate than men's, which are more geared toward action, such as going to a sports event together. Consequently, women see conflict as a threat to cherished connections. Like Mama and Kuif in the Arnhem colony, they avoid confrontations at all cost. Women are extremely good at this, as evident from the lasting bonds they enjoy. But the depth of their relationships also means that if a fight does erupt, they're unable to say "nothing personal." Everything is intensely personal. This makes stepping back from discord, once it has burst through the surface, harder than for men.

## GO-BETWEENS

Vernon, the alpha male of the San Diego bonobo colony, regularly chased a younger male, Kalind, into the dry moat. It was as if Vernon wanted Kalind out of the group. The young male always returned, however, climbing the rope that hung down into the moat, only to be chased back again. After sometimes as many as a dozen such incidents in a row, Vernon usually gave in. He would fondle Kalind's genitals or engage in a rough tickling game. Without such friendly contact, Kalind would not be allowed to return. So, after emerging from the moat, his first task was to hang around the boss and wait for a cordial signal.

But the more intense and dramatic reconciliations among bonobos were always among the females, who could be fighting one moment and GG-rubbing the next. Reconciliations inevitably had a sexual element, and the same behavior could also be used preemptively to forestall conflict. When Amy Parish watched food division at the San Diego Zoo, she found that females would approach the food, hooting loudly, and engage in sex before touching it. The very first response, then, was not to eat or fight over the food, but to engage in frantic physical contact that served to calm tempers and pave the way for sharing. This is known as a "celebration," even though as you watch bonobos the word "orgy" might seem more appropriate.

A telling incident happened at the same zoo when the bonobos had just received a lunch of hearts of celery, all of which had been claimed by the females. Amy was taking pictures and gestured so that the apes would look her way for a photograph. But Loretta, who had most of the food, must have thought that Amy was begging. Loretta ignored her for about ten minutes. Then she suddenly stood up, divided her celery, and threw half of it across the moat to this woman who so desperately wanted her attention. This shows how intensely the females had embraced Amy as one of them, something they never did with me since apes make precise gender distinctions among people. Amy later visited these same bonobo friends after a maternity leave. She wanted to show the apes her infant son. The oldest female briefly glanced at the human baby, then disappeared into an adjacent cage. Amy thought the female was upset, but she had only left to pick up her own newborn. She quickly returned to hold the ape baby up against the glass so that the two infants could look into each other's eyes.

Among chimpanzees, celebrations are incredibly noisy affairs. They occur at the zoo if caretakers approach with buckets full of food and in the field if prey has been captured. The chimpanzees collect in large huddles to embrace, touch, and kiss. As with the bonobos, the partying occurs before anyone has tasted the food. Celebrations involve abundant body contact, marking a transition to a tolerant atmosphere in which everyone will get a share. But I must say that the most joyful celebrations I've ever seen in chimps had nothing to do with food. They happened every spring at the Arnhem Zoo when the apes heard the exterior doors being opened for the first time. The chimps knew each and every door in the building by ear. Having been cooped up for five winter months in a heated hall, the apes were ready to relax in the grass. Within a second of hearing these doors, the colony would utter a deafening scream that seemed to be coming out of a single throat. Once outside, the noise would continue while the chimpanzees stood around on the island in small groups jumping and thumping each other excitedly on the back. The mood was decidedly festive, as if this was the first day of a new and better life. Their faces would regain color in the sun, and tensions would dissolve in the spring air.

Celebrations demonstrate the need for physical contact at moments of exhilaration. This need is typical of all primates and easy for us to understand. We reach out to each other when our sports team wins or when a student graduates but also at upsetting times, such as funerals or following a calamity. This need for body contact is hardwired. Some cultures foster distance between people, yet a society devoid of body contact would not be truly human.

Our fellow primates understand this need for contact, too. They not only seek it for themselves, but also foster contact between others if doing so might ameliorate a strained relationship. The simplest example is that of a juvenile female who has been mothering the infant of another female. When the infant suddenly starts to cry, the juvenile will hurry back to the mother, quickly handing the screaming bundle over, knowing that this is the quickest way to calm it down. More sophisticated fostering of contact can be observed when male chimpanzees fail to reconcile after a confrontation. They will sometimes sit a couple of yards apart as if waiting for their adversary to make the first move. The uneasiness between them is obvious from the way they look in all directions—the sky, the grass, their own body—while scrupulously avoiding eye contact. Such a deadlock may last for half an hour or more, but can be broken by a third party.

A female will approach one of the males, and after having groomed him for a while, will slowly walk toward the other. If the first male follows, he does so right behind her without ever looking at the other male. Sometimes the female looks around to check and may return to tug the arm of a reluctant male in order to make him follow. When the female sits down close to the second male, both males will groom her, one on each side, until she simply walks off and leaves them to groom each other. The males will pant, splutter, and smack more loudly than before the female's departure, sounds with which chimps indicate their enthusiasm for the grooming. Such go-between behavior, called "mediation," allows male rivals to approach each other without taking any initiative, without making eye contact, and perhaps without losing face.

Mediation promotes peace in the community by bringing the disputants together. Interestingly, it's only female chimps who mediate and only the oldest and highest ranking among them. This is not so surprising, because if a male were to approach one of the rivals he'd just be perceived as party to the conflict. Given the propensity of male chimps to form alliances, his presence cannot be neutral. On the other hand, if a young female, especially one with swollen genitals, were to approach one of the two males, this would be interpreted sexually and would also increase tensions. In the Arnhem colony, Mama was the mediator par excellence: no male would ignore her or carelessly start a fight that might incur her wrath.

In other colonies, too, the highest ranking female had the skill and authority to bring male combatants together. I've even seen situations in which other females seemed to encourage this, approaching the top female while looking around at males who were refusing to make up, as if trying to get something going that they knew they couldn't undertake themselves. In this sense, female apes clearly have the capacity for peacemaking, and a very advanced one at that. But notice that their mediations concern males. Males are amenable to such efforts, whereas females may not be. I have never seen a female trying to bring two female rivals together after a fight.

Humans, of course, can barely coexist without intermediaries. This holds true in any society, large or small. Harmonization of clashing interests is institutionalized and guided by social influences including the role of elders, foreign diplomacy, the court system, conciliatory feasts, and compensatory payments. The Malayan Semai, for example, hold a *becharaa'*—an assembly of the disputants, their relatives, and the rest of the community at the headman's house. The Semai know how much is at stake: they have a saying that there are more reasons to fear a dispute than a tiger. The *becharaa'* is opened by monologues in which the elders hold forth for many hours about mutual dependencies within the community and the need to maintain good relations. Disputes concern serious matters, such as infidelity and ownership, and are resolved in discussions that may last for days, in which the entire community goes over all possible motives of the contestants, the reasons why things happened, and how they could have been prevented. The session ends with the headman charging one or both disputants never to repeat what they've done since it might endanger everybody.

The common good is nothing to sneeze at. Or, as Keith Richards said to Mick Jagger when the Rolling Stones almost broke up, "This is bigger than the both of us, baby."

## THE ESCAPE GOAT

"Victory has a hundred fathers but defeat is an orphan," as the saying goes. Accepting responsibility for something that went wrong is not our strong point. In politics, we take the blame game for granted. Since no one wants it at his or her doorstep, blame tends to travel. This is the ugly way of resolving disputes: instead of reconciliation, celebration, and mediation, trouble arising at the top is dispatched to the bottom.

Every society has its scapegoats, but the most extreme cases I have known concerned newly established groups of macaques. These monkeys have strict hierarchies, and while the higher-ups were working out their rank positions, a process that tends to get nasty, nothing was easier for them than to turn en masse against a poor bottom-ranker. One female, named Black, got attacked so often that we called the corner she used to flee to "Black's corner." Black would crouch there while the rest of the group gathered around her, mostly grunting and threatening, but sometimes biting or pulling out fistfuls of hair.

In my experience of managing primates, there's no point in giving in to the temptation of removing the scapegoat: the next day another individual will take its place. There is an obvious need for a receptacle of tensions. But when Black gave birth to her first offspring everything changed, because the alpha male protected this infant. The rest of the group generalized their animosity to Black's family, hence they directed threats and grunts at this little monkey baby, too, but having high-level protection, he had nothing to fear and seemed rather puzzled by all the fuss. Black soon learned to keep her son close at times of trouble, because then no one would touch her either.

What makes scapegoating so effective is that it's a double-edged sword. First, it releases tension among the dominants. Attacking an innocent harmless bystander is obviously less risky than attacking each other. Second, it rallies the higher-ups around a common cause. While threatening the scapegoat, they bond with each other, sometimes mounting and embracing, indicating that they stand united. It's a total charade, of course: primates often pick enemies that hardly matter. In one monkey group, all the members would run to their water basin and threaten their own reflections. Unlike humans and apes, monkeys don't recognize their reflections as themselves, so they'd found an enemy group that conveniently didn't fight back. The Arnhem chimps had a different outlet. If tensions escalated to the breaking point, one of them would start barking at the lion and cheetah safari park next door. The big cats were perfect enemies. The entire colony would soon be wraah-barking at the top of its lungs at those awful beasts safely separated from them by a moat, a fence, and a strip of forest. Tensions would be forgotten.

A well-established group usually does not have a particular individual who is always chased into the corner. In fact, the absence of a whipping boy is a sure sign that things have settled down. But displacement of aggression, as it's known among specialists, does not necessarily end up at the bottom of the social ladder. Alpha threatens Beta, who immediately starts looking around for Gamma. Beta then threatens Gamma while glancing around at Alpha, because the ideal outcome is that Alpha takes Beta's side. Displacement aggression can trickle down four or five steps before it peters out. It's often of low intensity—the equivalent of name-calling or door-slamming—but still permits the higher-ups to let off steam. And everyone in the group knows what's going on: subordinates go into hiding at the first sign of tensions near the top.

The term "scapegoat" derives from the Old Testament, where it refers to one of two goats used in a ceremony on the Day of Atonement. The first goat was sacrificed, and the second was allowed to escape with its life (the "escape goat"). This goat received all inequities and transgressions of the people on its head before being sent out to a solitary land, which was literally the wilderness and symbolically a spiritual wilderness. This was the way people freed themselves from evil. Similarly, the New Testament describes Jesus as the "Lamb of God who takes away the sin of the world" (John 1:29).

For modern man, scapegoating refers to inappropriate demonization, vilification, accusation, and persecution. Humanity's most horrific scapegoating was the Holocaust, but letting off steam at the expense of others covers a far wider range of behavior, including witch-hunting in the Middle Ages, vandalism by the fans of losing sports teams, and spousal abuse after conflicts at work. And the mainstays of this behavior—the innocence of the victim and a violent release of tensions—are strikingly similar among humans and other animals. The quintessential example is pain-induced aggression in rats. Place two rats on an iron grid through which they are given an electric shock, and the moment they feel the pain they attack each other. Like people who hit their thumb with a hammer, the rats never hesitate to "fault" somebody else.

We surround this process with symbolism and pick victims based on things like skin color, religion, or a foreign accent. We also take care never to admit to the sham that scapegoating actually is. In this regard, we're more sophisticated than other animals. But it's undeniable that scapegoating is one of the most basic, most powerful, least conscious psychological reflexes of the human species, one shared with so many other animals that it may well be hardwired.

The mythical Oedipus died a scapegoat during social upheaval in his city, Thebes. Blamed for an enduring drought, he was the perfect victim, given that he was an outsider raised in Corinth. The same applies to Marie Antoinette. Political instability combined with her Austrian heritage made her an ideal target. Today, Microsoft is a scapegoat for the lack of internet security; illegal immigrants are held responsible for rising unemployment; and the Central Intelligence Agency took the fall for the never-found weapons of mass destruction in Iraq.

The Iraq war itself is another good example. Like all Americans, I was shocked and stunned by the terrorist attack on New York. In addition to my initial horror and grief, anger soon became part of the mix. I could feel it all around me, and I felt it percolating within myself as well. I'm not sure this feeling was shared by people in other parts of the world: horror and grief, yes, but anger, perhaps not. This might explain why what happened next put the United States at such monumental odds with other nations. Overnight, the world had to contend with a furious wounded bear, startled out of its slumber by someone who had stepped on its tail. As one popular song put it, a sucker punch had made the country light up like the Fourth of July.

After a swipe at Afghanistan, the angry bear kept looking for another, more substantial target, and there was Saddam Hussein, hated by everybody, most of all by his own people, thumbing his nose at the world. No matter the lack of any proven connection to 9/11, the bombing of Baghdad was a great tension release for the American people greeted by cheerleading media and flag-waving in the streets. Immediately following this catharsis, though, doubts began to surface. Eighteen months later, polls indicated that the majority of Americans considered the war a mistake.

Reallocation of blame fails to address the situation that triggered it, but it works. It serves to calm frayed nerves and restore sanity. As Yogi Berra put it, "I never blame myself when I'm not hitting. I just blame the bat." It's a good way of keeping yourself out of the equation, but how it exactly works is little understood. Only one study has measured it in a most innovative way: not in people, but in baboons. Primatologists have come up with "guidelines" for what makes for a successful male baboon. The measure of success is the amount of glucocorticoid (a stress hormone that reflects how one is doing psychologically) in the blood. Low levels mean that one copes well with the ups and downs of social life, which for male baboons is full of status striving, slights, and challenges. It was found that displacement of aggression is an excellent personality trait for a baboon to have. As soon as a male has lost a confrontation, he takes it out on some smaller guy. Males who tend to do so enjoy relatively stress-free lives. Rather than withdrawing and sulking after a defeat, they're quick to shift their problems to others.

I've heard women say that this is a male thing, that women tend to internalize blame, whereas men have no compunction about finding others at fault. Men prefer to give rather than get ulcers. It's depressing to learn that we share this tendency—which creates so many innocent victims—with rats, monkeys, and apes. It's a deeply ingrained tactic to keep stress at bay at the expense of fairness and justice.

## THIS CROWDED WORLD

As a young scientist, I once asked a world-famous expert on human violence what he knew about reconciliation. He gave me a lecture on how science should focus on the causes of aggression, since causes hold the key to elimination. My interest in conflict resolution suggested to him that I took aggression for granted, something he did not approve of. His attitude reminded me of opponents of sex education: Why waste time on improving behavior that should not even exist?

The natural sciences are more straightforward than the social sciences. No topic is taboo. If something exists and can be studied, it deserves to be studied. It's as simple as that. Reconciliation not only exists, it is extremely widespread among social animals. Quite in contrast to the violence expert, I feel that our only hope of curbing aggression lies in a better understanding of how we're naturally equipped to handle it. To focus attention exclusively on the problematic behavior is to be like a firefighter who learns everything about fire but nothing about water.

One of the triggers of aggression that scientists often mention is in fact relatively unimportant precisely because of the checks and balances that our species has in place—the link between crowding and aggression. The nineteenth-century English demographer Thomas Malthus noted that human population growth is automatically slowed by rising vice and misery. This inspired psychologist John Calhoun to conduct a nightmarish experiment. He put an expanding rat population in a crammed room and observed how the rats soon set about sexually assaulting, killing, and cannibalizing each other. As predicted by Malthus, population growth was naturally curbed. The chaos and behavioral deviancy led Calhoun to coin the phrase "behavioral sink." Normal rat behavior had gone down the drain, so to speak.

In no time, street gangs were being compared to rat packs, inner cities to behavioral sinks, and urban areas to zoos. We were warned that an ever more crowded world was heading for either anarchy or dictatorship. Unless we stopped breeding like rabbits, our fate was sealed. These views entered mainstream thinking to the point that you can ask almost anyone and they'll tell you that crowding is one of the main reasons we have trouble eliminating human violence.

Primate research initially supported this harrowing scenario. Scientists reported that city-dwelling monkeys in India were more aggressive than those in the forest. Others claimed that primates in zoos were excessively violent, ruled by bullies who dominated a social hierarchy that was an artifact of captivity: in the wild, peace and egalitarianism prevailed. Borrowing from the hyperbole of popularizers, one crowd study reported a "ghetto riot" among baboons.

While I worked with rhesus monkeys at the Henry Vilas Zoo in Madison, Wisconsin, we received complaints that the monkeys were fighting all the time, hence we must be housing too many of them together. To me, however, those macaques seemed perfectly normal: I had never known a group of rhesus monkeys that didn't squabble. In addition, having grown up in one of the most crowded nations in the world, I'm profoundly skeptical about any link between crowding and aggression. I just don't see it in human society. So, I devised a large-scale study of rhesus monkeys that had lived under particular circumstances for many years, often generations. The most crowded groups inhabited cages, and the least crowded lived on a large forested island. The island monkeys had six hundred times more space per capita than the caged ones.

Our first finding was that, surprisingly, density does not affect male aggressiveness in the least. In fact, the highest aggression rates were found in free-ranging males, not captive ones. Crowded males groom females more, and the females groom them more. Grooming has a calming effect: a monkey's heart rate slows down while being groomed. Females reacted differently. Rhesus females have a strong sense of belonging to a kin-clique, known as a matriline. Since these cliques are in competition with each other, crowding induces frictions. But not only does aggression go up between matrilines, as one might expect, grooming does as well. This means that females are working hard to keep tensions at bay by grooming outside their matrilines. As a result, the effect of crowding on monkeys is far less dramatic than one might think.

We speak of "coping," meaning that primates have ways of countering the effects of reduced space. Perhaps owing to their greater intelligence, chimpanzees go even further. I still remember a winter when the young upstart, Nikkie, seemed ready to challenge the current alpha male, Luit, in the Arnhem colony. The apes lived in an indoor hall in which confrontation with the established leader would amount to suicide. After all, Luit had extensive female support: the females would have helped him corner his adversary had Nikkie tried anything. But as soon as the colony was let outdoors, trouble started. Females move more slowly than males, and on the large island Nikkie could easily evade whatever defenses were mounted on Luit's behalf. In fact, all power struggles in Arnhem have occurred outdoors, never indoors. We know that chimps have a concept of the future, so we shouldn't put it past them to bide their time until conditions are favorable for stirring things up.

This kind of emotional control is also seen in conflict avoidance when chimps are housed in cramped quarters. Then they actually *reduce* aggression. They're a bit like people on an elevator or city bus, who ease frictions by minimizing large body movements, eye contact, and loud talking. These are small-scale adjustments, but it's also possible for entire cultures to adapt to the amount of space available. People in crowded countries often stress tranquility, harmony, deference, modulated voice levels, and respect of privacy even if walls are literally paper-thin.

Our sophisticated ability to adapt to a particular socio-ecology, as a biologist would call it, explains why the number of people per square mile has no bearing whatsoever on murder rates. Some nations with sky-high homicide rates, like Russia and Colombia, have very low population density, and among those with the lowest murder rates, we find Japan and the Netherlands, countries filled to the brim with people. This also applies to urban areas, where most crimes occur. The world's most densely packed metropolis is Tokyo, and one of the most sprawling is Los Angeles. Nevertheless, Los Angeles has about fifteen murders annually for every one hundred thousand people compared with Tokyo's under two.

In 1950, the world counted 2.5 billion people. We're now at around 6.5 billion. This is a steep climb since dating began, two millennia ago, when the human population of the entire world was estimated at between 200 and 400 million. If crowding indeed leads to aggression, we would be in for total combustion. Fortunately, we hail from a long line of social animals capable of adjusting to all sorts of conditions, including unnatural ones like crowded pens, city streets, and shopping malls. The adjustment may not be without effort, and the exuberant celebrations each spring at the Arnhem Zoo certainly indicate that chimps prefer a less crowded existence. But adjustment is preferable to the frightening alternative predicted on the basis of Calhoun's rat experiment.

I should add, though, that Calhoun's results may not entirely have been the product of crowding. Since the rats were given only a few food hoppers, competition probably played a role as well. This is a warning for our own species in an ever more populous world. We have a natural, underappreciated talent to handle crowding, but crowding combined with scarcity of resources is an entirely different story, one that might well lead to the vice and misery Malthus foresaw.

Malthus had an incredibly callous political outlook, though. He believed that any assistance given to the poor negates the natural process according to which these people are supposed to die off. If there was one right that man *didn't* possess, he said, it was a right to subsistence that he himself could not purchase. Malthus inspired a system of thought, known as Social Darwinism, devoid of compassion. Accordingly, self-interest is society's lifeblood, which translates into progress for the strong at the expense of the weak. This justification of disproportionate resources in the hands of a happy few was successfully exported to the New World, where it led John D. Rockefeller to portray the growth of a business as "merely the working-out of a law of nature and a law of God."

Given the popular use and abuse of evolutionary theory, it's hardly surprising that Darwinism and natural selection have become synonymous with unchecked competition. Darwin himself, however, was anything but a Social Darwinist. On the contrary, he believed there was room for kindness in both human nature and in the natural world. We urgently need this kindness, because the question facing a growing world population is not so much whether or not we can handle crowding, but if we will be fair and just in the distribution of resources. Will we go for all-out competition or will we do the *humane* thing? Our close relatives can teach us some important lessons here. They show us that compassion is not a recent weakness going against the grain of nature but a formidable power that is as much a part of who and what we are as the competitive tendencies it seeks to overcome.

FIVE

# KINDNESS

## Bodies with Moral Sentiments

Any animal whatever, endowed with well-marked social instincts . . . would inevitably acquire a moral sense or conscience, as soon as its intellectual powers had become as well developed, or nearly as well developed, as in man.

—CHARLES DARWIN

Why should our nastiness be the baggage of an apish past and our kindness uniquely human? Why should we not seek continuity with other animals for our "noble" traits as well?

—STEPHEN JAY GOULD

Eleven years had passed since I had last seen Lolita. I walked up to her cage, and as soon as I called her name, she rushed forward to greet me with pantgrunts, not a behavior chimps show to strangers. Of course, we remembered each other. When she still lived at the Yerkes Field Station, we saw each other every day and got along very well.

Lolita is special to me for one simple, charming act she once performed, which made clear how much apes are underestimated. It's hard to get a good look at a newborn ape, which is really no more than a little dark blob against a mother's dark tummy. But I was eager to see Lolita's baby, which had been born the day before. I called her out of the group and pointed at her belly. Lolita looked up at me, sat down, and took the infant's right hand in her right hand and its left hand in her left hand. This sounds simple, but given that the baby was clinging to her, she had to cross her arms to do so. The movement resembled that of people crossing their arms when grabbing a T-shirt by its hems in order to take it off. She then slowly lifted the baby into the air while turning it around on its axis, unfolding it in front of me. Suspended from its mother's hands, the baby now faced me instead of her. After the baby made a few grimaces and whimpers—infants hate losing touch with a warm belly—Lolita quickly tucked it back into her lap.

With this elegant little motion Lolita demonstrated that she realized I would find the face of her newborn more interesting than its back. To take someone else's perspective represents a huge leap in social evolution. Our golden rule—"Do unto others as you would have them do unto you"—asks us to put on someone else's shoes. We think of this as a uniquely human ability, but Lolita showed we are not alone. How many animals can do so? I have already described how Kuni, a bonobo, treated an injured bird she found in her enclosure. By trying to make the bird fly, Kuni recognized the needs of an animal totally unlike herself. There is no shortage of further examples of bonobos figuring out the needs of others.

One involves Kidogo, who suffered from a heart condition. He was feeble, lacking the normal stamina and self-confidence of a grown male bonobo. When first introduced to the colony at the Milwaukee County Zoo, Kidogo was completely confused by the keepers' shifting commands inside the unfamiliar building. He failed to understand where to go if people urged him to move from one part of the tunnel system to another. After a while, other bonobos stepped in. They approached Kidogo, took him by the hand, and led him to where the keepers wanted him, thus showing they understood both the keepers' intentions and Kidogo's problem. Soon Kidogo began to rely on their help. If he felt lost, he would utter distress calls, and others would quickly come over to calm him and act as a guide.

That animals help each other is far from a new observation, but it's puzzling nonetheless. If all that matters is survival of the fittest, shouldn't animals refrain from anything that fails to benefit themselves? Why help another get ahead? There are two main theories: First, that such behavior evolved to help kin and offspring, hence individuals who are genetically related. This promotes the helper's own genes as well. This "blood is thicker than water" theory explains, for example, the sacrifice of bees, who give their lives for their hive and queen when stinging an intruder. The second theory follows an "If you scratch my back, I'll scratch yours" logic: if animals help those who return the favor, both parties stand to gain. Mutual aid can explain political alliances, such as between Nikkie and Yeroen, who supported one another and shared the gains in power and sexual privileges.

Both theories concern the evolution of behavior, but neither tells us much about actual motives. Evolution depends on the success of a trait over millions of years; motives spring from the here and now. For example, sex serves reproduction, yet when animals couple, it's not out of a desire to reproduce. They don't know the connection: sexual urges are separate from the reason sex exists. Motivations lead a life of their own, which is why we describe them in terms of preferences, desires, and intentions rather than survival value.

Consider the zoo bonobos who helped Kidogo. Clearly, none of them was related to him, nor could any expect much help in return from a debilitated individual. Possibly, they just liked Kidogo, or felt for him. In the same way, Kuni expressed concern for a bird despite the fact that helping behavior surely did not evolve in bonobos for the benefit of nonbonobos. Once a tendency exists, however, it's permitted to soar free from its origin. In 2004, Jet, a black Labrador in Roseville, California, jumped in front of his best friend, a boy, who was about to be bitten by a rattlesnake, and took the serpent's venom. Jet was rightly considered a hero. He wasn't thinking of himself; he was a genuine altruist.

This shows the risks animals are prepared to take. The boy's grateful family spent four thousand dollars on blood transfusions and veterinary bills to save their pet. A zoo chimp was less fortunate as he gave his life in a failed attempt to rescue an infant of his species, which had fallen into the water due to its mother's clumsiness. Since apes can't swim, to enter water takes incredible courage.

Altruistic behavior is common in humans. Once a week, my Atlanta newspaper lists "random acts of kindness" in which people recount stories about the help they've received from strangers. One elderly woman wrote of the day her eighty-eight-year-old husband, who was on his way out of the house, found that a huge pine tree had fallen across their driveway. A stranger driving by jumped out of his pickup truck, cut up the tree with the chain saw he had with him, and moved it to the side, clearing the couple's driveway. When the woman went out to pay the man for his work, he was already gone.

Don't think helping strangers is always easy. When Lenny Skutnik dove into the icy Potomac River in 1982 to rescue a victim of a plane crash, or when European civilians sheltered Jewish families during World War II, they took incredible risks. During earthquakes people regularly run into collapsing or burning houses to drag strangers out of them. A reward may come afterward in the form of praise on the evening news, but this can never be the motive. No sane person would risk his life for a minute of televised glory. In the 9/11 chaos of New York City, numerous anonymous acts of heroism occurred.

But even though we and other social animals occasionally assist others without thinking of ourselves, I would still argue that these tendencies originate from mutuality and the assistance of kin. Jet, the hero dog, likely considered the boy a member of his pack. Early human societies must have been optimal breeding grounds for "survival of the kindest" aimed at family and potential reciprocators. Once this sensibility had come into existence, its range expanded. At some point, sympathy for others became a goal in itself: the centerpiece of human morality and an essential aspect of religion. Thus, Christianity urges us to love our neighbor as ourselves, clothe the naked, feed the poor, and tend the sick. It is good to realize, though, that in stressing kindness, religions are enforcing what is already part of our humanity. They are not turning human behavior around, only underlining preexisting capacities.

How could it be otherwise? One cannot sow the seeds of morality on unwilling soil any more than one can train a cat to fetch the newspaper.

## HOW EMPATHIC AN ANIMAL?

Once upon a time, the president of a large nation was known for a peculiar facial display. In an act of barely controlled emotion, he would bite his lower lip and tell his audience: "I feel your pain." Whether or not the display was sincere is hardly the issue—being affected by another's predicament is. Empathy and fellow-feeling are second nature to us, so much so that anyone devoid of it strikes us as mentally ill or dangerous.

At the movies, we can't help but inhabit the characters on the screen. We despair when we see them drown as their gigantic ship sinks; we exult when they stare into a long lost lover's eyes. There's not a dry eye in the room even though all we do is sit in a chair staring at a screen. We all know empathy, yet it took a long time before it was taken seriously as a research topic. Too soft for the taste of hard-nosed scientists, empathy used to be lumped in with telepathy and other supernatural phenomena.

Times have changed, and my chimps recently made the point during a visit by one of the pioneers of empathy research in children, Carolyn Zahn-Waxler. Carolyn and I went to see the Yerkes colony. Among the apes was a female named Thai who is extremely attracted to people. She is in fact more interested in us than she is in her fellow chimps. Each time I appear on the tower that overlooks the compound, she rushes forward with loud greeting grunts. I always greet her back and talk to her, after which she sits there staring at me until I leave.

This time, however, I was so engrossed in my discussion with Carolyn that I barely looked up. Since I had failed to acknowledge Thai, our discussion was interrupted by loud, high-pitched screams that grabbed our attention. Thai was hitting herself, as chimps do when they throw a tantrum, and was soon surrounded by others who put their arms around her, kissed her, or held her briefly in an attempt to reassure her. I immediately realized why she was making such a fuss and profusely greeted her, stretching a hand toward her from a distance. I explained to Carolyn that this chimp felt neglected because I had not said hello. Carolyn had no trouble recognizing the pattern. Thai kept looking at me with a big nervous grin on her face until she eventually calmed down.

The most interesting part of this incident was not that Thai took offense at my rudeness, but how the group reacted. This was exactly the sort of behavior Carolyn studies in children. Others tried to alleviate Thai's distress. In fact, Carolyn demonstrated this ability in animals even though animals were never her focus. When her team visited homes to find out how children respond to family members instructed to feign sadness (sobbing), pain (crying "ouch"), or distress (coughing and choking), they discovered that children a little over one year of age already comfort others. This is a milestone in their development: an aversive experience in someone they know draws out a concerned response, such as patting and rubbing the victim's injury. Because expressions of sympathy emerge in virtually all members of our species, they are as natural an achievement as the first step.

Not so long ago, it was assumed that empathy requires language. For some reason, a host of scientists see language as the source of human intelligence rather than its product. Since a one-year-old's behavior surely outstrips its verbal abilities, Carolyn's research showed that empathy develops well before language. This is relevant for animal research, which by necessity concerns nonverbal creatures. Her research team discovered that household pets, like dogs and cats, were as upset as children by distress-faking family members. The animals hovered over them, putting their heads in their laps with what looked like concern. Judged by the same standard as the children, the pets exhibited empathy as well.

Such behavior is even more striking in apes, where it has been dubbed "consolation." We measure consolation by simply waiting until a spontaneous fight occurs among our chimps after which we note if bystanders approach the victim. Bystanders often embrace and groom distressed parties. It is not unusual for a climbing youngster to fall out of a tree and scream. It will immediately be surrounded by others who hold and cradle it. This is exactly the response Binti Jua showed toward the boy at the Brookfield Zoo. If an adult loses a fight with a rival and sits screaming alone in a tree, others will climb toward him to touch and calm him. Consolation is one of the most common responses among apes. We recognize this behavior because apes do it in a way identical to ours, except for the occasional sexual consolation among bonobos.

The empathic response is one of the strongest there is, in fact stronger than the ape's proverbial desire for bananas. This was first reported by an early-twentieth-century Russian psychologist, Nadie Ladygina-Kohts, who raised a young chimpanzee, Yoni. Every day, Kohts had to deal with his unruly behavior. She discovered that the only way to get Yoni off the roof of her house was to appeal to his concern for her:

> *If I pretend to be crying, close my eyes, and weep, Yoni immediately stops his plays or any other activities, quickly runs over to me, all excited and shagged, from the most remote places in the house, such as the roof or the ceiling of his cage, from where I could not drive him down despite my persistent calls and entreaties. He hastily runs around me, as if looking for the offender; looking at my face, he tenderly takes my chin in his palm, lightly touches my face with his finger, as though trying to understand what is happening.*

At its simplest, empathy is the ability to be affected by the state of another individual or creature. This can be just body movement, such as when we mimic the behavior of others. We put our arms behind our head if others do the same and follow our colleagues at a meeting in crossing or uncrossing our legs, leaning forward or backward, adjusting our hair, putting elbows on the table, and so on. We do this unconsciously, especially with companions whom we like, which explains why couples who have lived together for a long time often resemble each other. Their demeanor and body language have converged. Knowing the power of body mimicry, researchers can manipulate people's feelings about each other. Being with someone who adopts deviating body postures—because she has been told to do so—results in fewer good feelings about her than being with someone who dutifully copies every move we make. When people say they "click," or are falling in love, they are unconsciously influenced by the amount of reflexive body mimicry they have engaged in as well as other subtle signs of openness to the other, such as keeping their legs apart or closed, raising or folding their arms, and so on.

As a child, I involuntarily mimicked body movements of others especially if I was actively involved, such as during sports. At some point I became aware of this, and I tried to suppress it, but couldn't. I have a photograph of myself during a volleyball game, where I jump up and act as if I am hitting a ball even though it is one of my brothers who actually has the ball. I am just acting out what I think he should be doing with it. This tendency is easily seen when human parents feed their young. While pushing a spoon full of gooey stuff toward the baby's mouth, adults open their *own* mouth when the baby is supposed to open his or hers, often followed by tongue movements simultaneous with the baby's. Similarly, when children are older and perform in a play at school, parents in the audience mouth the words their offspring are supposed to say.

Bodily identification is common in animals. A friend of mine once broke his right leg, which was put in a cast. Within days his dog started to limp, dragging her right leg. A veterinarian carefully checked the dog, but found nothing wrong. When, weeks later, the cast came off my friend's leg, the dog walked normally again. Similarly, in the Arnhem colony, Luit once injured a hand in a fight. He began supporting himself on a bent wrist, hobbling about in an odd manner. Soon thereafter, all juveniles in the colony began to walk in the same way. They kept this game up for months, long after Luit's injuries had healed. More immediate bodily identification has been described by Katy Payne for elephants: "Once I saw an elephant mother do a subtle trunk-and-foot dance as she, without advancing, watched her son chase a fleeing wildebeest. I have danced like that myself while watching my children's performances—and one of my children, I can't resist telling you, is a circus acrobat."

Monkeys scratch themselves if they see another do so, and apes yawn while watching a video of a yawning ape. We do the same, and not only in relation to our own kind. I once attended a slide show with photographs of yawning animals and found myself surrounded by an audience full of gaping mouths. I was quite unable to keep my own mouth closed. A research team at the University of Parma in Italy first reported that monkeys have special brain cells that become active not only if the monkey grasps an object with its hand but also if it merely watches another do so. Since these cells are activated as much by doing as by seeing someone else do, they are known as mirror neurons, or "monkey-see-monkey-do" neurons. Social animals relate to each other at a level far more basic than scientists previously suspected. We are hardwired to connect with those around us and to resonate with them, also emotionally. It's a fully automated process. Asked to watch photographs of facial expressions, we involuntarily copy the expressions seen. We do so even if the photo is shown subliminally, that is, if it appears for only a few milliseconds. Unaware of the expression, our facial muscles nevertheless echo it. We do the same in real life, as reflected in the classic Louis Armstrong lines "When you're smilin' . . . the whole world smiles with you."

Since imitation and empathy require neither language nor consciousness, we shouldn't be surprised that simple forms of relating to others exist in all sorts of animals, even the much-maligned rat. Already in 1959, a paper appeared with the provocative title "Emotional Reactions of Rats to the Pain of Others," which demonstrated that rats stop pressing a bar to obtain food if doing so delivers an electric shock to a rat next to them. Why didn't the rats simply continue to get food and ignore the other animal dancing in pain on an electric grid? In classic experiments (which I'd not wish to duplicate for ethical reasons), monkeys showed an even stronger inhibition. One monkey stopped responding for five days and another one for twelve days after witnessing a companion being shocked each time they pulled a handle to get food for themselves. These monkeys were literally *starving* themselves to avoid inflicting pain on others.

In all of these studies, the likely explanation is not concern about the other's welfare, but distress caused by another's distress. Such a response has enormous survival value. If others show fear and distress, there may be good reasons for you to be worried, too. If one bird in a flock on the ground suddenly takes off, all other birds will take off as well, before they even know what's going on. The one who stays behind may be prey. This is why panic spreads so quickly among people as well.

We have been programmed to thoroughly dislike seeing and hearing the pain of others. For example, young children often get teary-eyed and upset—and run back to their mothers for reassurance—when they see another child fall and cry. They are not worried about the other child, but overwhelmed by the emotions the other shows. It is only later in life, when children develop a distinction between self and other, that they separate vicarious emotions from their own. The development of empathy begins without any such distinction, however, perhaps similar to the way the vibrations of one string set off vibrations in another, producing a concerted sound. Emotions tend to arouse matching emotions, from laughing and joy to the well-known phenomenon of a room full of crying toddlers. We know now that emotional contagion resides in parts of the brain so ancient that we share them with animals as diverse as rats, dogs, elephants, and monkeys.

## IN THE OTHER'S SHOES

Every era offers humanity its own distinction. Seeing ourselves as special, we always look for confirmation. The first time this happened may have been Plato's definition of man as the only creature at once naked and walking on two legs. This sounded true enough until Diogenes arrived in the lecture hall with a plucked chicken, which he set loose with the words "Here is Plato's man." Thereafter, Plato's definition included "having broad nails."

Much later, the making of tools was regarded as so special that a book appeared under the title *Man the Tool-Maker.* This definition held until the discovery of wild chimps making sponges by chewing leaves into a wad or stripping leaves off branches before using them as sticks. Even crows have been seen to bend a metal wire into a hook so as to fish food out of a bottle. So much for man the tool-maker. The next claim was language, at first defined as symbolic communication. When linguists heard about apes with sign-language skills, however, they realized that the only way to keep these interlopers out was by dropping the symbol claim and stressing syntax instead. Humanity's special place in the cosmos is one of abandoned claims and moving goalposts.

The uniqueness claim du jour relates to empathy. It is not emotional connectedness per se, that is hard to deny in other animals, but a so-called theory-of-mind. This awkward phrase refers to the ability to recognize the mental states of others. If you and I meet at a party, and I believe that you believe that we have never met (even though I am sure we have), I have a theory about what is going on in your head. Taking another's perspective revolutionizes the way minds relate to each other. Given that some scientists claim this ability as uniquely human, it is ironic that the whole concept of theory-of-mind started with a 1970s primate study. Offered a choice among pictures, a chimp named Sarah would select a picture of a key if she saw a person struggle with a locked door or a picture of someone climbing onto a chair if she saw a man jump up and down to reach a banana. It was concluded that Sarah recognized the intentions of others.

Since this discovery, a whole industry of child research on theory-of-mind has sprung up, whereas primate research has had its ups and downs in this area. A few experiments on apes have failed, leading some to conclude that apes altogether lack theory-of-mind. Negative results are hard to interpret, though. As the saying goes, "Absence of evidence is no evidence of absence." When comparing apes and children, one problem is that the experimenter is invariably human, so that only the apes face a species barrier. And who says that apes believe that people are subject to the same laws as themselves? To them, we must seem from a different planet.

Recently, for example, my assistant called me about a fight in which Socko was wounded. The next day, I walked up to him and asked him to turn around, which he obligingly did—having known me since he was young—to show me his behind with the gash. Think about this from the ape's perspective. They are smart animals, always trying to figure out what is going on. Socko must have wondered how I knew about his injury.

If we come across as all-knowing gods, doesn't this make us unsuitable for experiments on the connection between seeing and knowing, which is the centerpiece of theory-of-mind? All that most of these experiments have done is test the ape's theory of the human mind.

We better focus on the ape's theory of the ape mind. When a creative student, Brian Hare, managed to cut out the human experimenter, he found that apes realize that if another has seen hidden food, this individual knows. Brian tested our chimps by enticing a low-ranking individual to pick up food in front of a higher ranking one. The subordinate ape went for pieces that the other could not have seen. In other words, chimps know what others know, and they use this information to their advantage. This threw the question of animal theory-of-mind wide open again. In an unexpected twist (because the debate revolves around humans and apes), a capuchin monkey at the University of Kyoto recently passed a number of seeing-knowing tasks with flying colors. A few such positive outcomes suffice to put a giant question mark behind previous negative ones.

This reminds me of a period in the almost century-long history of the Yerkes Primate Center when psychologists tried Skinnerian techniques on chimpanzees. One strategy was to deprive animals of food until they had dropped down to 80 percent of their body weight. This technique enhances motivation on food-oriented tasks in rats and pigeons. With apes, however, it failed to produce results. If anything, the apes became too morose and food-obsessed to pay attention to the job at hand. Primates need to enjoy something to be good at it. The harsh procedures of the rat psychologists created tensions at the center, including illicit feeding of apes by concerned personnel. When the investigators complained to the director that his chimps weren't nearly as smart as they'd been made out to be, the director blew up, famously reminding them that "there are no stupid animals, only inadequate experiments."

Exactly. The only way to get to the bottom of ape intelligence is to design experiments that engage them intellectually and emotionally. A few cups with hidden bits of food barely hold their attention. What they care about are social situations involving individuals close to them. Rescuing an infant from an attack, outmaneuvering a rival, avoiding conflict with the boss, and sneaking away with a mate are the kind of problems apes like to solve. The way Lolita turned her baby toward me, the way Kuni tried to rescue a bird, the way other bonobos led Kidogo by the hand, all suggest that real-life problems are sometimes solved by taking another's viewpoint. Even if each of these stories concerns only a single, unrepeated event, I attach great importance to them. Single events can be incredibly meaningful. After all, one step by one man on the moon has been sufficient for us to claim that going there is within our ability. If an experienced, reliable observer reports a remarkable incident, therefore, science better pay attention. And we do not have only one or two stories about how apes take another's perspective, but a great many of them. Let me offer a few more examples.

The two-meter-deep moat in front of the old bonobo enclosure at the San Diego Zoo had been drained for cleaning. After having scrubbed the moat and released the apes, the keepers went to turn on the valve to refill it with water when all of a sudden the old male, Kakowet, came to their window, screaming and frantically waving his arms so as to catch their attention. After so many years, he was familiar with the cleaning routine. As it turned out, several young bonobos had entered the dry moat but were unable to get out. The keepers provided a ladder. All bonobos got out except for the smallest one, who was pulled up by Kakowet himself.

This story matches my own moat observations, which concern the same enclosure a decade later. By this time, the zoo had wisely decided against water in the moat as apes cannot swim. A chain hung permanently down into it, and the bonobos visited the moat whenever they wanted. If the alpha male, Vernon, disappeared into the moat, however, a younger male, Kalind, sometimes quickly pulled up the chain. He would then look down at Vernon with an open-mouthed play face while slapping the side of the moat. This expression is the equivalent of human laughter: Kalind was making fun of the boss. On several occasions, the only other adult, Loretta, rushed to the scene to rescue her mate by dropping the chain back down and standing guard until he had gotten out.

Both observations tell us something about perspective taking. Kakowet seemed to realize that filling the moat while the juveniles were still in it wouldn't be a good idea even though this would obviously not have affected himself. Both Kalind and Loretta seemed to know what purpose the chain served for someone at the bottom of the moat and to act accordingly; the one by teasing, the other by assisting the dependent party.

During one winter at the Arnhem Zoo, after cleaning the hall and before releasing the chimps, the keepers hosed out all rubber tires in the enclosure and hung them one by one on a horizontal log extending from the climbing frame. Krom was interested in a tire in which there was still water. Unfortunately, this particular tire was at the end of the row, with six or more heavy tires hanging in front of it. Krom pulled and pulled at the one she wanted but couldn't remove it from the log. She pushed the tire backward, but there it hit the climbing frame and couldn't be removed either. Krom worked in vain on this problem for over ten minutes, ignored by everyone except Jakie, a seven-year-old Krom had taken care of as a juvenile.

Immediately after Krom gave up and walked away, Jakie approached the scene. Without hesitation he pushed the tires one by one off the log, beginning with the front one, followed by the second in the row, and so on, as any sensible chimp would. When he reached the last tire, he carefully removed it so that no water was lost, carried it straight to his aunt, and placed it upright in front of her. Krom accepted his present without any special acknowledgment and was already scooping up water with her hand when Jakie left.

That Jakie assisted his aunt is not so unusual. What is special, however, is that he, like Sarah in the original theory-of-mind experiments, correctly guessed what Krom was after. He grasped his aunt's goals. Such so-called targeted helping is typical of apes, but rare or absent in most other animals.

As we have seen with Kuni and the bird, apes care about other species. This may sound paradoxical given that chimps brutally kill and eat monkeys in the wild. But is it really that hard to understand? We are ambivalent, too. We love animals as pets, but we also slaughter them (sometimes the same animals). The fact that chimps sometimes respond positively to potential prey shouldn't surprise us, therefore. I once saw the entire Yerkes colony intently follow how staff caught an escaped rhesus monkey in the forest around their enclosure. Attempts to lure the monkey back to his compound had failed. The situation became hairy when he climbed a tree. I heard Bjorn, then still a juvenile, suddenly whimper while grabbing the hand of an older female next to him. Bjorn's distress coincided with the monkey's clinging to a lower branch of the tree: he had just been hit by a tranquilizer dart. People were waiting underneath the tree with a net.

Although this was not a situation Bjorn himself had ever been in, he appeared to identify with the monkey: he uttered another whimper at the exact moment the escapee fell into the net.

At emotionally meaningful moments, apes can put themselves into another's shoes. Few animals have this capacity. For example, all scientists who've set out to find consolation in monkeys have come up empty-handed. Collecting exactly the same data we have for chimps, they found nothing. Monkeys fail to provide reassurance even if their own offspring has been bitten. They do protect them, but show none of the cuddling and stroking with which an ape mother calms down an upset youngster. This makes ape behavior so much more humanlike. What is it that sets apes and humans apart? Part of the answer may be increased self-awareness, because a second contrast has been known for even longer than the one regarding consolation. Apes are the only primates apart from ourselves to recognize their own reflection. Self-recognition is tested by giving an individual unknowingly a dot of paint in a place, such as above the eyebrow, that's not directly visible. After this, one gives the individual a mirror. Guided by their reflection, apes rub the painted spot with their hand and inspect the fingers that touched it, thus recognizing that the colored dot in the mirror actually sits on themselves. Monkeys make no such connection.

Every morning, when we shave or put on makeup, we rely on this ability. To see our mirror image as ourselves is entirely logical to us, but it's not something we expect in another animal. Imagine your dog walking by the mirror in the hallway and screeching to a halt the way we do when something unusual catches our eye. We would be shocked! The dog cocks his head and checks out his image in the mirror, shaking his head to unfold a folded ear or removing a twig stuck in his fur. Dogs never do this, but it's exactly the sort of attention apes pay to themselves. If I approach my chimps with sunglasses on, as happens often in the summer, they make weird grimaces while staring into my glasses. They jerk their heads at me until I take the glasses off and hold them close to them as a mirror. Females turn around to look at their behinds—a logical obsession given the appeal of this body part—and most apes open their mouth to inspect the inside, touching their teeth with their tongue or picking at them with their fingers guided by the mirror. Sometimes, they go so far as to "embellish" themselves. Presented with a mirror, Suma, an orangutan at a German zoo, gathered salad and cabbage leaves from her cage, placed them on top of each other, then put the whole pile on her head. Staring in the mirror, Suma carefully rearranged her vegetable hat. One would swear she was getting ready for a wedding!

Self-awareness affects how we deal with others. Around the time children first recognize themselves in a mirror—at between eighteen and twenty-four months of age—they also develop helping geared to the needs of others. Their development parallels the transformation during our evolution: self-recognition and higher forms of empathy emerged together in the branch leading to humans and apes. A connection between these capacities was predicted decades ago by Gordon Gallup, the American psychologist who first used mirrors to test primates. Gallup assumed that empathy requires self-awareness. Perhaps it works as follows: in order to act on behalf of someone else, one needs to separate one's own emotions and situation from those of the other. The other needs to be seen as an independent entity. The same distinction between self and other permits us to recognize that our mirror image, which acts exactly like us, is *not* an independent entity. From this, we conclude that it must represent us.

When it comes to these capacities, we should not write off other animals, though. Many animals are extremely social and cooperative, making them excellent candidates for higher forms of empathy. Two species that come to mind are elephants and dolphins. Elephants are known to use their trunk and tusks to lift up weak or fallen comrades. They also utter reassurance rumbles to distressed juveniles. Dolphins have been known to save companions by biting through harpoon lines, hauling them out of tuna nets in which they got entangled, and supporting sick companions close to the surface so as to keep them from drowning. They assist people in the same way, such as in a recent account of four swimmers off the New Zealand coast whom dolphins herded away from a nine-foot shark.

Similarities with the consolation and targeted helping of apes beg the question of how elephants and dolphins respond to mirrors. Is there a parallel here as well? For elephants, this question remains open, but it can hardly be coincidental that the only nonprimate for which there is any evidence for mirror self-recognition is the dolphin. When bottle-nosed dolphins at the New York Aquarium were marked with painted dots, they spent more time in front of a mirror than if they had been left unmarked. The first thing they did upon arriving at the mirror (which was at a distance from where the marking took place) was spin around so as to get a good look at their markings.

Empathy is widespread among animals. It runs from body mimicry—yawning when others yawn—to emotional contagion in which the self resonates with fear or joy when it picks up fear or joy in others. At the highest level we find sympathy and targeted helping. Perhaps empathy has reached its peak in our species, but several other animals—most notably apes, dolphins, and elephants—come close. These animals understand the other's predicament well enough to offer optimal assistance. They drop a chain to those who need to climb up, support those who need a gulp of air, and take a disoriented party by the hand.

They may not know the golden rule, but they surely seem to follow it.

## SPOCK'S WORLD

KIRK: You'd make a splendid computer, Mr. Spock.
SPOCK: That's very *kind* of you, Captain!

Imagine a world filled with creatures like the super-logical Mr. Spock of *Star Trek*. If emotions did occasionally make an appearance, no one would know what to make of them. Being sensitive to language content only, they would miss changes in tone of voice and never engage in the human equivalent of grooming: small talk. Lacking any natural connection to each other, the only way these creatures could figure each other out would be through an arduous process of asking and learning.

Focusing exclusively on the dog-eat-dog side of evolution, an entire literature has depicted us as inhabitants of Spock's autistic universe. Kindness, we are told, is something people engage in only under pressure, and morality is little else than a veneer, a thin overlay hiding our selfish nature. But who actually lives in such a world? A bunch of piranhas driven to kindness because they want to impress each other would never develop the sort of societies we depend on. Unconcerned about each other, piranhas lack morality as we know it.

Mutual dependence is key. Human societies are support systems within which weakness does not automatically spell death. The philosopher Alasdair MacIntyre opens his book *Dependent Rational Animals* by pointing out the extent of human vulnerability. During many life stages, especially when we are young and old, but also in between, we find ourselves in the caring hands of others. We are inherently needy. So why do Western religion and philosophy pay so much more attention to the soul than the body? They depict us as cerebral, rational, and in charge of our destinies; never sick, hungry, or lusty. That humans have bodies and emotions is acknowledged only as a weakness.

During a public debate about the future of humanity, a respected scientist once ventured that in a couple of centuries we would gain full scientific control over our emotions. He seemed to be looking forward to the day! Without emotions, however, we would barely know which life choices to make, because choices are based on preferences, and preferences are ultimately emotional. Without emotions we wouldn't store memories, because it's the emotions that make them salient. Without emotions we would remain unmoved by others, who in turn would remain unmoved by us. We would be like ships sailing past each other.

The reality is that we are bodies born from other bodies, bodies feeding other bodies, bodies having sex with other bodies, bodies seeking a shoulder to lean or cry on, bodies traveling long distances to be close to other bodies, and so on. Would life be worth living without these connections and the emotions they arouse? How happy would we be, especially given that happiness, too, is an emotion?

We have become forgetful, according to MacIntyre, of how much our basic concerns are those of an animal. We celebrate rationality, but when push comes to shove we assign it little weight. As any parent who has tried to talk sense into a teenager knows, the persuasive power of logic is surprisingly limited. This is especially true in the moral domain. Imagine an extraterrestrial consultant who instructs us to kill people as soon as they come down with the flu. In doing so, we are told, we would kill far fewer people than would die if the epidemic were allowed to run its course. By nipping the flu in the bud, we would save lives. Logical as this may sound, I doubt that many of us would opt for this plan. This is because human morality is firmly anchored in the social emotions, with empathy at its core. Emotions are our compass. We have strong inhibitions against killing members of our own community, and our moral decisions reflect these feelings.

Empathy is intensely interpersonal. It is activated by the presence, demeanor, and voice of others rather than by any objective evaluation. Reading about the plight of someone who has fallen on hard times is really not the same as sharing a room with this person and listening to his story. The former situation may generate some empathy, but it's of a type that's easily ignored. Why? For rational moral agents, the two situations shouldn't make a difference. But our moral tendencies evolved in direct interaction with others whom we could hear, see, touch, and smell, and whose situation we understood by taking part in it. We're exquisitely attuned to the stream of emotional signals coming from other people's faces and postures, and we resonate with expressions of our own. Actual people get under our skin in a way that an abstract problem never will. The English concept of "empathy" derives from the German *Einfühlung,* which translates as "feeling into."

My flu example goes to show that we refuse to strive for the greatest good for the greatest number of people (a school of moral philosophy known as "utilitarianism") if doing so violates the basic inhibitions of our species. The other approach, which is Immanuel Kant's claim that we arrive at morality by "pure reason," poses even greater problems. This was explored by a young philosopher with neuroscience interests, Joshua Greene, who scanned people's brains while asking them to solve moral dilemmas. One dilemma went as follows: You are at the wheel of a trolley that lacks brakes. You are speeding toward a fork in the track and notice five rail workers on the left fork and only one on the right fork. All you can do is choose the trolley's path by throwing a switch. There is no time to brake. What would you do?

The answer is simple. Most people will turn right so that they kill only one worker. But what if you're standing on a bridge overlooking a straight track that has no fork and on which you see a trolley bearing down full speed on five rail workers. A heavyset man is standing next to you. You could push him off the bridge. He would drop in front of the trolley, slowing it down enough so that all workers would be saved. As it turns out, people are far more willing to kill a person by changing the trolley's direction than by willfully sending someone to his death. This choice has little to do with rationality, because logically the two solutions are the same: five persons are saved at the expense of one. Kant wouldn't have seen any difference.

We have a long evolutionary history in which seizing someone with our bare hands had immediate consequences for ourselves and our group. Bodies matter, which is why anything relating to them arouses emotions. In the scanner, Greene discovered that moral decisions, such as whether or not to push someone off a bridge, light up areas of the brain that serve both the person's own emotions and the evaluation of other people's emotions.

Impersonal moral decisions, in contrast, which evolution has not prepared us for, activate areas that we also use for practical decisions. Throwing the switch on the trolley is treated by our brain like any neutral problem, such as the question of what we will eat today or at what time we need to leave home to make our flight.

Moral decision-making is driven by emotions. It activates parts of the brain that go back to the transition from cold-blooded reptiles to the nursing, caring, loving mammals that we are. We are equipped with an internal compass that tells us how we ought to treat others. Rationalizations often come after the fact, when we have already carried out the preordained reactions of our species. Perhaps rationalization is a way of justifying our actions to others, who then may agree or disagree so that society as a whole can reach consensus about a certain moral dilemma. This is where social pressure comes in—the approval and disapproval that are so important to us—but all of this is probably secondary to "gut level" morality.

This may be shocking to the Kantian philosopher, but it fits Charles Darwin's conviction that ethics grew out of social instincts. Following in Darwin's footsteps, Edward Westermarck, a Swedish-Finnish anthropologist writing at the beginning of the twentieth century, understood how little control we exert over our moral choices. Rather than being the product of reasoning, Westermarck noted, "we approve and disapprove because we cannot do otherwise. Can we help feeling pain when the fire burns us? Can we help sympathizing with our friends? Are these phenomena less necessary or less powerful in their consequences, because they fall within the subjective sphere of experience?"

Before Darwin and Westermarck, similar ideas had been expressed by David Hume, the Scottish philosopher who stressed moral sentiments, and well before all of them there was the Chinese sage Mencius (372–289 B.C.), a follower of Confucius. Recorded on bamboo clappers handed down to his descendants, his writings show how little is new under the sun. Mencius believed that people tend toward the good as naturally as water flows downhill. This is evident from his remark about our inability to bear the suffering of others:

> *If men suddenly see a child about to fall into a well, they will without exception experience a feeling of alarm and distress. They will feel so, not as a ground on which they may gain the favor of the child's parents, nor as a ground on which they may seek the praise of their neighbors and friends, nor from a dislike to the reputation of having been unmoved by such a thing. From this case we may perceive that the feeling of commiseration is essential to man.*

Remarkably, all possible selfish motives listed by Mencius (such as wanting favors and seeking praise) are detailed at length in the modern literature. The difference is, of course, that Mencius rejected these explanations as too contrived, given the immediacy and force of the sympathetic impulse. Manipulation of public opinion is entirely possible at other times, Mencius said, but not at the moment a child is about to fall into a well.

I couldn't agree more. Evolution has equipped us with genuinely cooperative impulses and inhibitions against acts that might harm the group on which we depend. We do apply these impulses selectively, but are affected by them nonetheless. I don't know if people are deep down good or evil, but I do know that, despite his impressive intelligence, Mr. Spock wouldn't be able to solve moral dilemmas in a manner that would satisfy us. He would go about it too logically. Pushing that man off the bridge, he would be baffled by the victim's protestations and our revulsion.

## GENEROSITY PAYS

One balmy evening at the Arnhem Zoo, when the keeper called the chimps inside two adolescent females refused to enter the building. The weather was superb. They had the whole island to themselves and they loved it. The rule at the zoo was that none of the apes would get fed until all of them had moved inside. The obstinate teenagers caused a grumpy mood among the rest. When they finally did come in, several hours late, they were assigned a separate bedroom by the keeper so as to prevent reprisals. This protected them only temporarily, though. The next morning, out on the island, the entire colony vented its frustration about the delayed meal by a mass pursuit ending in a physical beating of the culprits. That evening, they were the first to come in.

Punishment of transgressors relates to the second pillar of morality, which concerns resources. The young females had caused every stomach in the colony to rumble. We are talking bodies again, but now in a different way. Stomachs need regular filling. The result is competition. Having or not having, appropriating, stealing, reciprocity, fairness: all have to do with the division of resources, a top concern of human morality.

But perhaps I have a peculiar view of morality, which I should explain. For me morality has to do with either Helping or (not) Hurting. The two H's are interconnected. If you are drowning and I withhold assistance, I am in effect hurting you. My decision to help or not is by all accounts a moral one. Anything unrelated to the two H's, even though presented as a moral issue, falls outside morality. It's most likely a mere convention. For example, one of my first culture shocks when I moved to the United States was hearing that a woman had been arrested for breast-feeding in a shopping mall. It puzzled me that this could be seen as offensive. My local newspaper described her arrest in moral terms, something having to do with public decency. But since natural maternal behavior cannot conceivably hurt anybody, it was no more than a norm violation. By the age of two, children distinguish between moral principles ("do not steal") and cultural norms ("no pajamas at school"). They realize that breaking some rules harms others, but breaking other rules just violates expectations. The latter kind of rules are culturally variable. In Europe, no one blinks an eye at naked breasts, which can be seen at every beach, but if I were to say I had a gun at home, everyone would be terribly upset and wonder what had become of me. One culture fears guns more than breasts, while another fears breasts more than guns. Conventions are often surrounded with the solemn language of morality, but in fact they have little to do with it.

The critical resources relating to the two H's are food and mates, which are both subject to rules of possession and exchange. Food is most important for female primates, especially when they are pregnant or lactating (which they are much of the time), and mates are most important for males, whose reproduction depends on the number of fertilized females. It is logical, therefore, that sex-for-food deals among apes—in which copulation leads to the sharing of food—are asymmetrical: males go for the sex, females for the food. Since the giving and receiving occur almost at the same time, these deals are a simple form of reciprocity. True reciprocity is quite a bit more complicated. We often do favors that are repaid days or months later, meaning that we rely on trust, memory, gratitude, and felt obligations. This is so much part of our society that we would be shocked if someone failed to grasp the idea of reciprocity.

Let us say that I helped you move a piano down the narrow stairs in your apartment building. Three months later, I'm moving myself. I call you to explain that I have a piano, too. If you wave me off with "Good luck with it!" I may remind you of what I did for you, even though this would be aggravating. If you still don't offer any help, I may explicitly mention the idea of tit-for-tat. I would find this most embarrassing. But if your response is "Oh, but I don't believe in reciprocity!" this would be truly disturbing. It would be an out-and-out negation of why we humans live in groups, of why we do each other any favors at all. Who would ever want to deal with you? Even if we understand that repaying a favor is not always possible (for example, if you have to be out of town the day I move, or if you have a bad back), it's hard to understand anybody who openly denies quid pro quo. The denial makes you an outcast: someone lacking a crucial moral tendency.

When Confucius was asked if there is a single word that could serve as a prescription for all of one's life, he arrived, after a lengthy pause, at "reciprocity." This elegant, all-encompassing principle is a human universal, and biologists have a long-standing interest in its origins. I still remember the excitement when, in 1972, with a number of students at the University of Utrecht, we analyzed Robert Trivers's "The Evolution of Reciprocal Altruism." It remains one of my favorite articles because rather than simplifying the connection between genes and behavior, it pays full attention to emotions and psychological processes. It distinguishes different types of cooperation based on what each participant puts into and gets out of it. For example, cooperation with immediate rewards does not qualify as reciprocal altruism. If a dozen pelicans form a semicircle in a shallow lake to herd small fish with their paddling feet, all birds profit when they scoop up the prey together. Because of the instant payoffs, this kind of cooperation is widespread. Reciprocal altruism, on the other hand, costs something before it delivers. This is more complicated.

When Yeroen supported Nikkie's bid for dominance, he couldn't know if this would be successful. It was a gamble. When Nikkie did reach the top, however, Yeroen immediately made his desires plain, trying to mate with females under Nikkie's nose. For obvious reasons, no other male tried this, but Nikkie was dependent on the old male's support, so had to let him. This is classical reciprocity: a transaction that serves both parties. After having analyzed thousands of alliances in which individuals support each other in fights, we concluded that chimps reach high levels of reciprocity. That is, they support those who support them.

They also reciprocate in the negative sense: they know revenge. Revenge is the flip side of reciprocity. Nikkie had a habit of squaring accounts not long after his occasional defeat by an alliance. He would corner a group mate who had participated in the alliance when he or she was alone. With the other allies out of sight, this individual would be in for a terrible time. As a result, every choice has multiple consequences, both good and bad. It is obviously risky for a low-ranking individual to get even with a high-ranking one, but if the latter is already under attack, there could be an opening to make him or her bleed. Payback is just a matter of time. By the end of my stint at the Arnhem Zoo, I was so attuned to the dynamics within the colony that I could predict who would jump in when and how. I would watch a female, Tepel, who had been seriously injured earlier in the week by another female, Jimmie, to see what Tepel would do when Jimmie was on the receiving end of a fight with Mama, the uncontested queen. As expected, Tepel saw this as a golden opportunity to add her two cents to Jimmie's defeat, thus reminding her to pick her enemies with care.

Another female, Puist, once took the trouble to help her male friend, Luit, chase off Nikkie. When Nikkie afterward did what he so often did, single out Puist for reprisals, Puist naturally turned to Luit, who was nearby, stretching out her hand for support. Luit didn't lift a finger to protect her, though. Immediately after Nikkie had left the scene, Puist turned on Luit, barking furiously. She chased him across the enclosure. If her fury was in fact the result of her friend's failure to help her after she had helped him, the incident suggests that chimpanzee reciprocity is ruled by expectations similar to those among humans.

An easy way to find out about reciprocity is to exploit the fact that chimps share food. In the wild, they chase monkeys until they capture one, then they tear it apart so that everyone gets a piece. The hunt I saw in the Mahale Mountains followed this pattern, with males begging around the monkey carcass high up in the trees. The male who had captured the meat held on to it, but at some point he gave half to his best buddy, who immediately became the center of a second begging cluster. It took two hours, but by the end virtually everyone in the tree owned a piece. Females with swollen genitals were more successful than other females in getting food. And it's known that among the males, hunters favor fellow hunters when dividing the meat. Even the most dominant male, if he hasn't been part of the hunting party, may stay empty-handed. This is another example of reciprocity: those who contributed to success have priority in the division of spoils.

Food sharing likely started as an incentive for hunters to hunt another day: there can't be joint hunting without joint payoffs.

One of my favorite Gary Larson cartoons shows a group of primitive men, shovels in hand, returning from the forest carrying a giant carrot above their heads. The text reads "Early vegetarians returning from the kill." The carrot was large enough to serve the whole clan. This is profoundly ironic given how unlikely it is that vegetables played any role in the evolution of food sharing. The leaves and fruits that primates collect in the forest are too abundant and too small to share. Sharing makes sense only in relation to highly prized food that is hard to obtain and comes in amounts too large for a single individual. What is the centerpiece when people gather around the dinner table? The turkey at Thanksgiving, the pig turning on the spit, or the salad bowl? Sharing goes back to our hunting days, which explains why it is rare in other primates. The three primates best at public sharing—that is, sharing outside the family—are humans, chimpanzees, and capuchin monkeys. All three love meat, they hunt in groups, and they share even among adult males, which makes sense given that males do most of the hunting.

If a taste for meat is indeed at the root of sharing, it is hard to escape the conclusion that human morality is steeped in blood. When we give money to begging strangers, ship food to starving masses, or vote for measures that benefit the poor, we follow impulses shaped since our ancestors first gathered around a meat possessor. At the center of the original circle is something desired by many but obtainable only with exceptional strength or skill.

Food sharing lends itself perfectly to research on reciprocity. Instead of patiently waiting for spontaneous events, I simply hand food to one of my chimps and follow the trickle-down economy until all the others have gotten their share. This procedure lets me determine who has what for sale on the "market of services," which covers political support, protection, grooming, food, sex, reassurance, and myriad other favors. (Of course, I am not so cruel as to give my chimps live prey, although they sometimes trap a raccoon or cat at the field station. They never eat these, however, since they are well fed and lack hunting traditions.) We offer them watermelons or a tight bundle of branches with leaves—large enough to share, but also easily monopolized. Sharing didn't originate in relation to such foods, but now that the tendency exists, we can measure it during vegetarian meals.

When we carry food toward the colony, the chimpanzees burst out in a "celebration" during which they kiss and embrace each other. This takes a couple of minutes, after which I drop the bundle of branches from the tower toward May, for instance. May, a low-ranking female, will look around before picking it up. If Socko approaches at the same time, she won't touch the food. Instead, she'll step back and let him have it. But if she arrives first and has her hands on the food, it's hers. This is noteworthy, because people think that dominant individuals can claim everything. Not so in chimpanzees. Jane Goodall related with some astonishment how her most dominant male had to beg for his share. This goes by the name of "respect of possession." It doesn't apply to juveniles, who lose their food rather quickly, but even the lowest-ranking adult can keep his or her portion without being bothered. My explanation has again to do with reciprocity. If Socko were to steal May's food, there would be little she could do about it. But the event would be filed away in May's head. This wouldn't be to Socko's advantage, since there are many services over which he has no control. If he were to offend females in the group by being a bully, where would he be if he got into trouble with a rival, if he needed grooming, wanted someone to lick his wounds, or desired sex? In a market of services, everyone has leverage.

Most sharing occurs in a remarkably calm atmosphere. Ape beggars hold out their hand, palm upward, very much as human beggars do on the street. They whimper and whine, but confrontations are rare. These occur if a possessor wants someone to leave the circle and whacks him or her over the head with her food or barks at him or her in a shrill voice until her or she leaves. Food is obtained through tolerance. Beggars gingerly reach for a leaf, and if they meet no resistance, try something more brazen, such as removing an entire branch or breaking one off. Friends and family of the possessor are least hesitant. May is one of our more generous sharers. She sometimes keeps the best branches (such as blackberry shoots and sassafras) for herself, but gives the rest away. This isn't because of her low status. Another low-ranking female, Georgia, is so stingy that no one even bothers begging from her. Georgia tends to keep everything for herself. This makes her unpopular. If she herself wants food from someone, she needs to solicit for the longest time. If May wants food, on the other hand, she just moves right into a circle and starts feeding. This is the beauty of reciprocity: generosity pays.

For our project, we measured grooming among the chimpanzees in the morning to compare it with feeding in the afternoon. A vast number of observations allowed us to relate success in obtaining food with previous grooming. If Socko had groomed May, for example, his chances of getting a few branches from her went up markedly compared to days on which he had not groomed her. Ours was the first animal study to statistically demonstrate an exchange of favors after an interval of several hours. Moreover, the exchanges were partner-specific; that is, May's tolerance specifically benefited Socko, the one who had groomed her, and no one else.

Since we ourselves behave much the same way, these results sound obvious enough. But consider the abilities involved. One is memory of previous events. This is no big deal for chimpanzees, who can remember a face for over a decade. May simply needed to remember Socko's grooming. The second ability is to color this memory so that it triggers friendly feelings. In ourselves, we call this coloring process "gratitude," and chimps seem to have the same ability. Whether they also feel obligations is unclear, but it's interesting that the tendency to return food for grooming is not the same in all relationships. For individuals who associate a lot, such as May and her friends or daughters, a single grooming session barely causes a ripple. All sorts of daily exchanges occur between them, probably without their keeping track. It's all part of their bond. It is only in the more distant relationships, such as between May and Socko, that a grooming session stands out and is specifically rewarded.

It's hardly different in our case. Over dinner at a workshop on social reciprocity, one of the experts confided that he kept track on his computer every day of what he had done for his wife and what she had done for him. Lots of forks wavered between plates and mouths as we tried to come to grips with what we had just learned. The consensus was that this was a bad idea, that if you keep close track with friends, let alone a spouse, there is probably something amiss. The man was talking about his third wife and now lives with his fifth, so perhaps we were onto something. Favors are exchanged almost unthinkingly in close relationships. Generally, these relationships are highly reciprocal, but they also have room for temporary imbalances, sometimes permanent ones, such as when a friend or a spouse falls ill. It is only in the more distant relationships that account sheets are kept. Similar to chimps, we repay an acquaintance or colleague who has shown unexpected kindness but we would not necessarily repay our best friend. The latter's help is appreciated, too, but is part of a deeper, more fluid relationship.

Like clerks doing the bookkeeping of a harbor, we are aware of all incoming and outgoing goods and services. We repay received help with help and received hurt with hurt, keeping the two H's on roughly equal footing with those around us. We do not like unnecessary imbalances. The same aversion explains why the two teenage chimps were disciplined: They had upset too many balances at once. They needed to be taught a lesson—the one Confucius saw as the greatest in life.

## ETERNAL GRATITUDE

Mark Twain once quipped, "If you pick up a starving dog and make him prosperous, he will not bite you. This is the principal difference between a dog and a man."

Knocking human shortcomings is fun, especially in comparison with animal behavior. There may, in fact, be some truth to Twain's assessment. In my home, we have adopted stray pets, and I must say that they seem eternally grateful. A scrawny kitten full of fleas, picked up in San Diego, grew into a gorgeous tomcat by the name of Diego. Diego purred excessively his entire fifteen-year life whenever he was fed—even on occasions when he barely ate anything. He seemed more appreciative than most pets, perhaps because he had experienced an empty stomach in his youth. I am not sure that we should call this "gratitude," though. It could have been mere happiness. Instead of realizing that he owed his comfortable life to us, Diego may just have enjoyed food more than your average spoiled pet.

But now take the following ape story. Two chimps had been shut out of their shelter during a rainstorm. Wolfgang Köhler, the German pioneer of tool-use studies, happened to walk by and found the apes soaking wet, shivering in the rain. He opened the door for them. But instead of hurrying past him to enter the dry area, both chimps hugged the professor in a frenzy of satisfaction. This comes much closer to gratitude.

I have my own experience in this regard, which brings me back to Kuif and Roosje, whose introduction to the Arnhem colony was described in the first chapter. We had two reasons for giving Roosje to Kuif for adoption. Roosje had been born to a deaf female, Krom. We didn't want Krom to raise any more offspring after previous ones had died. Mother apes depend on small sounds of contentment and discomfort from their babies to know how they're doing. If Krom sat down on her baby, however, she wouldn't even notice its screams. The feedback chain was broken. We removed Roosje on the fourth day after birth. Instead of giving her to a human family—a common solution—we decided to keep her in the colony. Young apes raised in human homes become too human-oriented and lack skills to get along with other apes. Kuif was the perfect candidate for an adoptive mother. She had lost some of her own offspring due to insufficient lactation and so had no offspring to compete with Roosje. Kuif was extremely interested in ape babies, however. In fact, we had noticed that if Krom neglected her infant's cries, Kuif sometimes started crying, too.

Each time one of her own offspring had died, Kuif had gone into a deep depression marked by rocking, self-clutching, refusing food, and heart-wrenching screams. When we trained her to bottle-feed Roosje, the infant remained firmly in our possession, however, even though Kuif desperately wanted to hold it. The training must have been rather frustrating, because Kuif was not allowed to drink from the bottle herself: she was asked to insert it through the bars for Roosje, who remained on our side. After weeks of training, Kuif performed these actions to our satisfaction and we made the transfer, placing the wriggling infant in the straw of Kuif's night cage. At first, Kuif stared closely into Roosje's face without touching her: in her mind, the baby belonged to us. Taking the baby of someone else without permission isn't well-regarded among chimps. Kuif approached the bars where the keeper and I sat watching. She kissed both of us, glancing between Roosje and us as if asking permission. We urged her, waving our arms at the infant, saying "Go, pick her up!" Eventually, she did, and from that moment on Kuif was the most caring and protective mother one could imagine, raising Roosje as we'd hoped.

Kuif's reintroduction to the colony several months later didn't go without a glitch. Not only did we have Nikkie's hostility to deal with, but Roosje's natural mother was upset. A couple of times, Krom tried to snatch Roosje from Kuif, something I had never seen a female do before, or since. But since Kuif was higher-ranking, she was able to defend herself, and Mama helped out, too. Is it possible that Krom still recognized Roosje even though she hadn't seen her since the removal? I was skeptical about this explanation until I heard about a human mother who recognized her own daughter although she hadn't seen her since she was a baby. It happened in Philadelphia in 2004. This woman's ten-day-old daughter had disappeared in a fire. The mother had never accepted her child's death, however. She had found an open window in her burned home and was convinced there had been a break-in. Many years later, the mother went to a children's party where she spotted a girl she immediately thought was her daughter. She managed to pull a few strands of hair from the child. The mother and six-year-old daughter were later reunited based on DNA evidence from the hair sample. A neighbor admitted to having stolen the baby and setting the house on fire to cover her tracks.

This remarkable case of identification (the mother said she recognized the "dimples" in the child's cheeks) is just an aside to show how well mothers study their babies. For the same reason, Krom may have sensed who Roosje was. My main point, however, is the effect the adoption had on Kuif's relationship with me, which brings me back to gratitude. The two of us had a rather neutral relationship before, but ever since the adoption—now almost three decades ago—Kuif showers me with the utmost affection whenever I show my face. No other ape in the world reacts to me as if I am a long-lost family member, wanting to hold my hands and whimpering if I try to leave. Our training allowed Kuif to raise not only Roosje, but some of her own children as well, since she'd been giving them the bottle, too. She has been eternally appreciative.

Gratitude is concerned with balance sheets. It makes us help those who have helped us. This must be its original function, even though now we apply the feeling more widely, being grateful for splendid weather or good health, for instance. That gratitude is a virtue may explain why it gets so much more attention than its ugly sister, revenge. Revenge is concerned with balance sheets, too, but for the other of the two H's. Bitterness toward those who hurt us is common, and here, too, feelings are translated into actions, such as the settling of scores. Not only do we feel vengeful ourselves, we worry about these feelings in those whom we have offended, knowing how every chicken comes home to roost. We know the mechanism so well that we may in fact propose revenge upon ourselves, seeing the acceptance of punishment as the only way to restore the peace.

Let me illustrate this with an example from opera, which is a spectacle I like to watch and listen to when I am not watching primates. With its dramatized human relations, opera depicts the side of human behavior that philosophers often ignore and social scientists often consider secondary to our acclaimed rationality. But human life, or at least the part that matters most to us, is thoroughly emotional. Apart from love, solace, guilt, hate, jealousy, and so on, opera never has any shortage of revenge nor of the sweet satisfaction it is known to deliver.

Vendetta is the grand theme of Mozart's *Don Giovanni*, in which the vile protagonist, after a life of seduction and deception, meets an angry mob and eventually his maker. In a side scene, a peasant woman, Zerlina, nearly falls for Don Giovanni on the day of her wedding to Masetto. Zerlina comes back to her angry husband with a great need for forgiveness. In an aria titled "Batti, batti, o bel Masetto" ("Beat Me, Beat Me, Oh Handsome Masetto") the guilty woman begs for punishment. She promises that she will sit like a lamb while Masetto tears out her hair, digs out her eyes, and beats her with his bare hands. She knows that the only way to make up is to let her husband square the account, meaning that some of his hurt be transferred to herself. She promises to kiss the hands that strike her. Not a politically correct message perhaps, but then the attraction of opera is its display of raw emotions. Masetto loves Zerlina too much to take her up on the offer, however, and all ends well.

Westermarck saw retribution as the centerpiece of human morality, and he thought we were not the only species to display it. In his time, there was little research on animal behavior, so he had to rely on anecdotes, such as the one he heard in Morocco about a vengeful camel. The camel had been excessively beaten by a fourteen-year-old boy for turning the wrong way. The animal passively took his punishment, but a few days later, finding itself unladen and alone on the road with the same conductor, "the camel seized the unlucky boy's head in its monstrous mouth, and lifting him up in the air flung him down again on the earth with the upper part of the skull completely torn off, and his brains scattered on the ground."

Stories of resentful animals can be heard at many a zoo, usually concerning elephants (with their proverbial memories) and apes. Every new student or caretaker working with apes needs to be told that they won't be able to get away with pestering or insulting them. The ape remembers and will take all the time in the world to get even. Sometimes it does not take long. One day, a woman came to the Arnhem Zoo's front desk complaining that her son had been hit by a big rock thrown by the chimpanzees. The son was surprisingly subdued, however. Witnesses later said he'd thrown the same rock at the chimpanzees first.

Our data indicate that chimps also square accounts among themselves. When they take sides in a confrontation between others, they go against those who often go against themselves. It's impossible to experiment on this topic without inciting nasty behavior, however. That's why we test only the positive side of reciprocity, such as in our work on capuchin monkeys. Capuchins are quite different from apes. They are little brown monkeys, about the size of a cat, with long tails that serve as marvelous grasping organs. They are native to Central and South America, meaning that they evolved separately from our own African stock for at least thirty million years. They are about the smartest monkeys I know. Sometimes called New World chimpanzees, their brain, relative to their body, is as large as that of an ape. Capuchins use tools, have complex male politics, have lethal encounters between groups, and most important, share food. This makes them ideal for studies of reciprocity and economic decision-making.

Our capuchins are divided into two groups, in which they breed, play, fight, and groom. They live in enclosures with an outdoor area, but are trained to individually enter a test chamber for brief periods. The tests, which involve food, are so much to their liking that they compete over who gets to go. Most of the time, we set up a test in the chamber, turn on a video camera, and watch the monkeys on a screen in the next-door office. In a typical experiment we put two monkeys side by side, offering Individual A a bowl of cucumber slices, after which we give Individual B a bowl of apple pieces. We then measure how they share with each other. Since we put mesh between the monkeys, they cannot steal each other's food. They are forced to wait for whatever the other brings close to the partition, allowing them to reach for it. Most primates would stay in their own little corner and jealously keep all food away. But this is not the capuchin way. Our monkeys bring lots of food to where the other can get at it and occasionally even push it through the mesh to the other.

We found that if Individual A had been generous with its cucumber, Individual B was more likely to share its apple. This was so encouraging that we turned the setup into a labor market. In a labor market, you pay me for the work that I do for you. We mimicked this by placing food on a movable tray with a counterweight too heavy for a single monkey to pull. Each monkey sat in its own side of the test chamber, ready to pull on a bar connected with the tray. Being true cooperators, they coordinated their actions perfectly to pull in the tray. The trick was that we placed the food in front of only one of the two monkeys, meaning that all benefits went to this individual, the winner. The other, the laborer, was just there to help. The only way the laborer could get anything was if the winner shared through the mesh.

Winners were more generous after cooperation than if they had obtained food by themselves. They seemed to realize when help was needed and rewarded those who provided it.

## FAIR IS FAIR

Where would we be without payment for labor? That this principle shows up in a monkey lab may seem surprising, but not if one knows that wild capuchins hunt giant squirrels. To catch such agile prey, which weigh up to one-quarter of the average capuchin male, is very difficult in the three-dimensional space of the forest, not unlike the monkey-hunting by chimpanzees. Unable to catch a squirrel on their own, capuchin hunters need help. Our experiment replicated the central issue of such cooperation, which is that it should provide payoffs not just to the one making the capture, but to everyone involved. Winners who fail to share won't be getting much help in the future, as we also noticed in our bar-pulling monkeys.

But what about the division of the spoils? That winners need to compensate laborers does not mean they have to give up everything. How much can they keep without upsetting the others? People are keenly aware of resource distribution, such as the difference between getting a small or large portion at the dinner table. The classic television sitcom *The Honeymooners* poked fun at this problem when Ralph Kramden (the fat one), Ed Norton (the skinny one), and their wives decided to share an apartment and eat together:

RALPH: When she put two potatoes on the table, one big one and one small one, you immediately took the big one without asking what I wanted.
NORTON: What would you have done?
RALPH: I would have taken the small one, of course.
NORTON: You would? (in disbelief)
RALPH: Yes, I would!
NORTON: So, what are you complaining about? You *got* the little one!

It all boils down to fairness. This is in fact a moral issue—relating to the two H's—though it isn't always presented as such. In the United States, it's not unusual for CEOs to earn a thousand times what the average worker makes. Those CEOs may be hurting others by claiming a disproportionate share of the corporate pie, yet they themselves will describe this as their privilege or as "the way the market works." Social Darwinism is trotted out to justify inequity, calling it only natural that some get more than others. Never mind that real Darwinism has a more subtle take on the division of resources. This is because we are cooperative animals in which even the mighty—perhaps especially the mighty—depend on others. The issue came up when Richard Grasso, chairman of the New York Stock Exchange, revealed a pay package for himself of close to 200 million dollars. There was public outcry about such outlandish compensation. As it happened, on the very day that Grasso was forced to resign, my team published a study on monkey fairness. Commentators couldn't resist contrasting Grasso with our capuchins, suggesting he could have learned a thing or two from them.

Sarah Brosnan and I tested fairness in a simple game. If one gives a capuchin a small pebble, then holds up something more appealing, such as a slice of cucumber, it grasps quickly that it needs to hand back the pebble to get the food. Capuchins have no trouble learning this game, since they naturally give and receive. Once they knew how to exchange pebbles for food, Sarah and I introduced inequity.

We placed two monkeys side by side, exchanging with both of them twenty-five times in a row, first with one, then the other, and so on. If both received cucumber, this was called equity. In this situation, the monkeys exchanged all the time, happily eating the food. But if we gave one of them grapes while keeping the other on cucumber, things took an unexpected turn. This was called inequity. Our monkeys' food preferences perfectly match the prices in the supermarket, so grapes are among the best rewards. Upon noticing their partner's salary raise, monkeys who had been perfectly willing to work for cucumber suddenly went on strike. Not only did they perform reluctantly but they got agitated, hurling the pebbles and sometimes even the cucumber slices out of the test chamber. A food they normally never refuse had become less than desirable: it had become disgusting!

This surely was a strong reaction equivalent to what, with some pomposity, is known in people as "inequity aversion." Admittedly, our monkeys showed an egocentric form of it. Rather than supporting the noble principle of fairness for everybody, they got upset about getting shortchanged. If general fairness had been their concern, advantaged monkeys should have shared an occasional grape with the other or refused grapes altogether, something they never did. The lucky grape recipients sometimes even supplemented their meal with their neighbor's abandoned cucumber slices. They were in a cheerful mood, as opposed to their poor partners who by the end of the test would sit sulking in the corner.

When Sarah and I published this study under the title "Monkeys Reject Unequal Pay" it struck a chord, perhaps because many people see themselves as cuke-eaters in a world with lots of grapes. All of us know how it feels to get the short end of the stick, which is why no parent would dare come home with a present for one child and not the other. An entire school of economics is convinced that emotions—which economists curiously call "passions"—play a major role in human decision-making. The most forceful ones are related to resource division. These emotions propel us to actions that seem irrational at first—such as quitting a job because we are paid less than others—but that in the long run promote a level playing field and cooperative relations.

This is tested with the so-called ultimatum game, in which one person receives, say, one hundred dollars to split with a partner. The split can be fifty-fifty, but it can also be any other division, such as ninety-ten. If the partner accepts the deal, both get their money. If the partner refuses, both end up with nothing. The one who divides the money needs to be careful, because partners commonly reject lowball offers. That they do so contradicts traditional economic theory according to which people are rational optimizers. A rational optimizer, however, should accept *any* offer, because even the smallest sum is better than no money at all. People don't think this way: they simply do not want to be taken advantage of. Grasso clearly had underestimated this sentiment.

Our monkeys showed the same reaction—throwing perfectly good food away! Cucumber is fine if there's nothing else, but as soon as others are munching on grapes, low-sugar veggies plunge in value.

## COMMUNITY VALUE

An egocentric sense of fairness is a fancy description of envy. It is the pain felt at the sight of those better off than we are. This is a far cry from the larger sense of fairness, the one that makes us also worry about those worse off than we are. If monkeys lack this latter sense, what about the apes? When I asked a fellow primatologist, Sue Savage-Rumbaugh, who conducts language studies on bonobos, for illustrations of empathy, she described what seems to me this larger sense.

Sue took care of a female, Panbanisha, while the rest of her bonobo colony was being tended by other staff. Panbanisha was getting different food, such as raisins and extra milk. As Sue brought these goodies to her, the other bonobos saw what was happening and called out. They obviously wanted the same stuff. Noticing this, Panbanisha seemed troubled, even though the situation was in her favor. She asked for juice, but when it arrived, instead of accepting it, she gestured to the others, waving an arm in her friends' direction and vocalizing at them. They responded with their own calls and then sat down next to Panbanisha's cage, waiting to get juice, too. Sue said she had the distinct impression that Panbanisha wanted her to bring the others what she herself was getting.

This is not enough to conclude that a sense of fairness exists in other animals, but what fascinates me is the connection with resentment. All one needs for the larger sense of fairness to develop is *anticipation* of the resentment of *others*. There are excellent reasons to avoid arousing bad feelings. Someone failing to share is excluded from feeding clusters. At worst, the one being envied risks being beaten up. Was this why Panbanisha avoided conspicuous consumption in front of her friends? If so, we are getting close to what may be the source of the fairness principle: conflict avoidance. It recalls the story of three boys in Amsterdam who fluttered two one-hundred guilder banknotes into the canal—a sum beyond a child's comprehension—because they had found five such notes. Since they couldn't share five notes equitably, they had decided to stay on good terms by getting rid of what they couldn't divide.

From humble beginnings noble principles arise. It starts with resentment if you get less, then moves to concern about how others will react if you get more, and ends with declaring inequity a bad thing in general. Thus, the sense of fairness is born. I like these step-by-step progressions, because this is how evolution must have worked. Similarly, we can see how revenge may, via intermediate steps, lead to justice. The an-eye-for-an-eye mentality of primates serves "educational" purposes by attaching costs to undesirable behavior. Although the human court system abhors raw emotions, there is no denying their role in our systems of justice. In her book *Wild Justice,* Susan Jacoby explains how justice is built upon the transformation of vengeance. When relatives of a murder victim or war survivors seek justice, they are driven by a need for redress even though they may present their cause more abstractly. Jacoby believes that one measure of a civilization's sophistication is the distance between aggrieved individuals and satisfaction of their urge for vindication, noting that there is "persistent tension between uncontrolled vengeance as destroyer and controlled vengeance as an unavoidable component of justice."

Personal emotions are crucial. Combined with an appreciation of how our behavior affects others, they create moral principles. This is the bottom-up approach: from emotion to a sense of fairness. It is quite the opposite of the view that fairness was an idea introduced by wise men (founding fathers, revolutionaries, philosophers) after a lifetime of pondering right, wrong, and our place in the cosmos. Top-down approaches (looking for an explanation by starting at the end product and working backward) are almost always wrong. They ask why we are the only ones to possess fairness, justice, politics, morality, and so on when the real question is what the building blocks are. What are the basic elements needed to construct fairness, justice, politics, morality, and so on? How did the larger phenomenon derive from simpler ones? As soon as one ponders this question, it is obvious that we share many building blocks with other species. Nothing we do is truly unique.

The big question of human morality is how we moved from interpersonal relations to a system that focuses on the greater good. I'm sure it isn't because we have the good of society foremost in mind. The first interest of every individual isn't the group, but itself and its immediate kin. But with increasing social integration, shared interests rose to the surface so that the community as a whole became a concern. We can see the beginnings of this when apes soothe relations between others. They broker reconciliations (bring parties together after a fight) and break up fights in an evenhanded manner in order to promote peace around them. This is because everybody has a stake in a cooperative atmosphere.

It's fascinating to see this sense of community at work, such as when Jimoh was stopped from pursuing a younger male by the barks of the entire chimpanzee colony. It was as if he hit a wall of dissent. Another incident in the same colony is still vivid in my mind. It centered on Peony, an older female. After the death of the group's alpha female, we didn't know for about one year which female was boss. Normally, it's one of the oldest females, so our bets were on three females who ranged between thirty and thirty-five years of age. Unlike males, females rarely show open rivalry over the top spot.

One day, I watched from my office as a small scuffle among a few young females began to involve adult males and ended up sounding extremely serious. The chimpanzees were screaming so loudly, and the males moving so fast, that I was pretty sure there would be blood in the end. But suddenly all commotion stopped and the males sat down, panting heavily, while several females hung around them. The atmosphere was extremely tense, and it was clear that nothing had been settled. They were just taking a break. It was at this moment that I saw who was alpha female. Peony got up from the tire in which she had been resting, and literally all eyes turned toward her. Some younger individuals walked up to her and some adults grunted softly the way they do to alert others, while Peony slowly and deliberately walked toward the center of the scene followed by everyone who had remained on the fringes. It looked like a procession, as if the queen had come to mingle with the commoners. All that Peony did was groom one of the two males, and soon other individuals followed her example, grooming each other. The second male joined the grooming clique as well. Calm returned. It was as if no one dared to start things up again after Peony had so gently put an end to it.

Peony, whom we informally call our "grooming machine," because she spends so much time grooming everyone, solves all problems by being sweet and calm and reassuring, which may explain why I didn't notice her being alpha. Subsequently, I've seen other incidents that confirm her absolutely central position, very much like Mama's in Arnhem but without the iron fist. It's at such moments that we realize that a chimp group is a true community rather than a bunch of individuals loosely hanging together.

Obviously, the most potent force to bring out a community sense is enmity with outsiders. It forces unity among elements normally at odds. This may not be visible at the zoo, but it's definitely a factor for chimps in the wild. In our own species, nothing is more obvious than that we band together against adversaries. This is why it's often suggested that the best guarantee for world peace would be an extraterrestrial enemy. We could finally put the Orwellian "war is peace" rhetoric into practice. In the course of human evolution, out-group hostility enhanced in-group solidarity to the point that morality emerged. Instead of merely ameliorating relations around us, as apes do, we have explicit teachings about the value of the community and the precedence it ought to take over individual interests.

And so the profound irony is that our noblest achievement—morality—has evolutionary ties to our basest behavior—warfare. The sense of community required by the former was provided by the latter. When we passed the tipping point between conflicting individual interests and shared interests, we ratcheted up the social pressure to make sure everyone contributed to the common good. We developed an incentive structure of approval and punishment—including internalized punishments such as guilt and shame—to encourage what's right and discourage what's wrong for the community. Morality became our main tool to strengthen the social fabric.

That the common good never stretched beyond the group explains why moral rules rarely mention the outside: people feel permitted to treat enemies in ways unimaginable within their own community. Applying morality beyond those boundaries is the great challenge of our time. In developing universal human rights—that must apply even to our enemies, as intended by the Geneva Convention—or debating the ethics of animal use, we apply a system that evolved for within-group reasons outside the group, even outside the species. Expansion of the moral circle is a fragile enterprise. Our best hope for success is based on the moral emotions, because emotions are disobedient. In principle, empathy can override every rule about how to treat others. When Oskar Schindler kept Jews out of concentration camps during World War II, for example, he was under clear orders by his society on how to treat these people, yet his feelings interfered.

Caring emotions may lead to subversive acts, such as the case of a prison guard who during wartime was directed to feed his charges only water and bread, but who occasionally sneaked in a hard-boiled egg. However small his gesture, it etched itself into the prisoners' memories as a sign that not all of their enemies were monsters. And then there are the many acts of omission, such as when soldiers could have killed captives without negative repercussions but decided not to do so. In war, restraint can be a form of compassion.

Emotions trump rules. This is why, when speaking of moral role models, we talk of their hearts, not their brains (even if, as any neuroscientist will point out, the heart as the seat of emotions is an outdated notion). We rely more on what we feel than what we think when solving moral dilemmas.

Mr. Spock's cerebral approach is woefully inadequate. This idea is superbly expressed in the parable of the Good Samaritan, which concerns our attitude toward people in need. A half-dead victim lies by the side of the road from Jerusalem to Jericho. The victim is ignored first by a priest, then by a Levite—both religious men familiar with the small print of everything written on ethics. These men didn't want to interrupt their journey for a victim they didn't know, and they quickly passed by on the other side of the road. Only the third passerby, a Samaritan, stopped, bandaged the man's wounds, put him on his donkey, and brought him to safety. The Samaritan, a religious outcast, felt compassion. The biblical message is to be wary of ethics by the book rather than by the heart and to treat everyone as your neighbor.

With morality firmly rooted in sentiment it's easy to agree with Darwin and Westermarck on its evolution and to disagree with those who think culture and religion contain the answer. Modern religions are only a few thousand years old. It's hard to imagine that human psychology was radically different before religions arose. It's not that religion and culture don't have a role to play, but the building blocks of morality clearly predate humanity. We recognize them in our primate relatives, with empathy being most conspicuous in the bonobo and reciprocity in the chimpanzee. Moral rules tell us when and how to apply these tendencies, but the tendencies themselves have been in the works since time immemorial.

SIX

# THE BIPOLAR APE

## On Striking a Balance

Are we best characterized by hate or by love? What is most critical for survival: competition or cooperation? Are we more like chimpanzees or bonobos?

Such questions are a waste of time for the bipolar characters that we are. It's like asking if a surface is best measured by its length or by its width. Even worse are attempts to consider only one pole at the expense of the other. Nevertheless, this is what the West has been doing for centuries by depicting our competitive side as somehow more authentic than our social one. But if people are as selfish as is assumed, how do they form societies? The traditional view is that of a contract among our ancestors, who decided to live together "by covenant only, which is artificial," as Thomas Hobbes put it. We are seen as loners who reluctantly joined forces: smart enough to pool resources, but lacking any true attraction to our own kind.

The old Roman proverb "Homo homini lupus" ("man is wolf to man") captures this asocial vision that continues to inspire law, economics, and political science today. The problem isn't just that this saying misrepresents us, but it also insults one of the most gregarious and loyal cooperators in the animal kingdom—so loyal, in fact, that our ancestors wisely domesticated them. Wolves survive by bringing down prey larger than themselves—such as caribou or moose—which they do through teamwork. Upon return from the hunt, they regurgitate meat for nursing mothers, the young, and sometimes the sick and old who stayed behind. Like chanting soccer fans, they reinforce pack unity by howling together before and after the hunt. There's no lack of competition, but wolves can't afford to let it run its course. Loyalty and trust come first. Behavior that undermines the bedrock of cooperation is dampened to prevent a meltdown of the social harmony upon which survival depends. A wolf who would let narrow individual interests prevail would soon find himself alone chasing mice.

Apes know the same solidarity. In a study in Taï National Park in Ivory Coast, chimpanzees took care of group mates wounded by leopards, licking away their blood, carefully removing dirt, and preventing flies from coming near the wounds. They waved the flies away, protected injured companions, and traveled slowly if these couldn't keep up. All of this makes perfect sense given that chimpanzees live in groups for a reason, the same way wolves and people are group-animals for a reason. We would not be where we are today had our ancestors been socially aloof.

What I see, therefore, is the opposite of the traditional image of a nature "red in tooth and claw," in which the individual comes first and society is a mere afterthought. One can't reap the benefits of group life without contributing to it. Every social animal strikes its own balance between the two. Some are relatively nasty, others relatively nice. But even the harshest societies, such as those of baboons and macaques, limit internal strife. People often imagine that in nature weakness automatically means elimination—a principle hyped as the "law of the jungle"—but in reality social animals enjoy considerable tolerance and support. Otherwise, what would be the point of living together?

I used to work with a group of rhesus macaques that showed great acceptance of Azalea, a mentally retarded juvenile born in their midst. Since Azalea possessed a triplet of chromosomes, her condition resembled that of human Down's syndrome. Rhesus monkeys normally punish anyone who violates the rules of their strict society but Azalea was able to get away with the gravest blunders, such as threatening the alpha male. It was as if everyone realized that nothing they did would ever change her ineptness. Similarly, a free-ranging macaque group in the Japanese Alps included a congenitally handicapped female named Mozu, who could barely walk, and certainly not climb, as she lacked both hands and feet. A frequent star of Japanese nature documentaries, Mozu was fully accepted by her group to the point that she lived a long life and raised five offspring.

So much for survival of the fittest. There is plenty of that, too, of course, but there's no need to caricature the life of our relatives as one led while constantly looking over their shoulders. Primates find great comfort in each other's company. Getting along with others is a critical skill because survival chances outside the group—where there are predators and hostile neighbors—are dismal. Primates who find themselves alone meet a quick death. This explains why they spend an extraordinary amount of time—up to 10 percent of the day—servicing social ties by grooming others. Field studies have shown that female monkeys with the best social connections have the most surviving offspring.

## AUTIST MEETS GORILLA

Bonding is so fundamental that an American woman with Asperger's syndrome, a form of autism, who had failed to come to terms with her condition while living among people, found inner peace after she began taking care of zoo gorillas. Or perhaps it was the gorillas who took care of her.

Dawn Prince-Hughes describes how people would unnerve her with their direct stares and direct questions to which they wanted immediate answers. The gorillas, on the other hand, left her space, avoided eye contact, and conveyed a comforting calm. Most of all, they were patient. Gorillas are "oblique" characters in that they rarely engage in face-to-face contact. Moreover, like all the apes, they lack the white sclera around the iris that makes the human gaze such an unsettling signal. Our eye coloring enhances communication, yet it also prevents subtleties of communication available to apes with their completely dark eyes. Moreover, apes rarely stare the way we do: they glance. They have incredible peripheral vision and follow much around them from the corners of their eyes. It's something to get used to. How often have I thought that apes failed to pay attention, only to be proven wrong? They hadn't missed a thing.

The way the gorillas empathized with Prince-Hughes by "looking without looking, and understanding without speaking," as she called it, took place via postures and body mimicry—the ancient animal language of connection. The colony's mighty silverback, Congo, was the most sensitive and comforting, directly responding to signs of distress. This comes as no surprise given that the male gorilla, despite his ferocious King Kong reputation, is a born protector. The horrifying tales of gorilla attacks with which colonial hunters used to come home were designed to impress us with human rather than gorilla bravery. But in fact, a charging gorilla male is prepared to die for his family.

That it takes an autistic person—someone considered deficient in interpersonal skills—to pick up on the primacy of ape bonding, and the deep kinship we sense with hairy bodies similar to ours, is remarkable. That Prince-Hughes was pulled out of her solitude by gorillas rather than by chimpanzees or bonobos makes sense in view of gorilla temperament. These apes are not nearly as extroverted as chimps and bonobos. Take the following ordeal that a Swiss zoo went through. One night, its chimps managed to remove the skylight of their indoor housing and escape to the roof, after which some of them traveled through the city, jumping from house to house. It took days to retrieve the apes, and the zoo was lucky that none of them was shot by the police or electrocuted by a power line.

This event gave a local animal rights group the idea of "liberating" the same zoo's gorillas. Without giving much thought to what might be best for the animals, they climbed onto the ape house at night and removed a skylight above the gorilla quarters. But even though the apes had many hours to escape, they didn't. The next morning, the caretakers found all of them sitting where they usually sat, staring with amazement up into the air, fascinated by the gaping hole above them. None of them had had the curiosity to climb out, and the skylight was simply put back into place. This, in a nutshell, is the temperamental difference between chimps and gorillas.

Our lineage's state of nature is one of bonding and support to the degree that even someone with autism perceives it. Or perhaps precisely such a person, since our obsession with the spoken word stands in the way of a full appreciation of nonverbal cues, such as postures, gestures, expressions, and tone of voice. Without body cues our communication loses its emotional content and becomes mere technical information. We could just as easily use flashing cards with "I love you" or "I am angry." It is well-known that people whose faces lose their expressiveness because of some neurological disorder and who therefore cannot echo the emotions of others (such as by a smile or frown), plunge into abject loneliness. Our species finds life barely worth living without the body language that cements us together.

Origin stories that neglect this deep connection by presenting humans as loners who grudgingly came together are ignorant of primate evolution. We belong to a category of animals known among zoologists as "obligatorily gregarious," meaning that we have no option but to stick together. This is why fear of ostracism lurks in the corners of every human mind: being expelled is the worst thing that can befall us. It was so in biblical times, and it remains so today. Evolution has instilled a need to belong and to feel accepted. We are social to our core.

## TAMED CONTRADICTIONS

A twenty-year-old Dodge Dart that I once owned taught me that the most critical part of a car is not its engine. The car would stop completely only if I applied my full weight to the brake pedal. On a quiet morning with little traffic, I drove it slowly to a nearby garage. Even though I arrived safely, the trip was frightening and I dreamed for months afterward of driving a car that would slow but not stop.

Nature's checks and balances are as essential as a car's brakes. Everything is regulated, all is kept in control. Mammals and birds, for example, have made the evolutionary leap toward warm-bloodedness, yet run into trouble each time they overheat. In hot weather or after exercise, they cool off by sweating, flapping their ears, or panting heavily with their tongue hanging out. Nature had to put a brake on body temperature. Similarly, every songbird has its optimum egg size, optimum clutch size, optimum travel distance for foraging, optimum prey size, and so on. Birds that deviate from the norm by laying too many eggs or looking for insects too far from the nest lose out in the evolutionary race.

This also applies to conflicting social tendencies, such as competition and cooperation, selfishness and sociability, strife and harmony. Everything is balanced around an optimum. Being selfish is inevitable and necessary, but only up to a point. This is what I meant when I called human nature a Janus head: we are the product of opposing forces, such as the need to think of our own interests and the need to get along. If I emphasize the latter, it's because of the traditional emphasis on the former. Both are closely interconnected and contribute to survival. The very capacities that promote peace, such as reconciliation after a fight, would never have evolved in the absence of conflict. In a bipolar world, every capacity hints at its very opposite.

We have discussed specific paradoxes, such as the link between democracy and hierarchy, between the nuclear family and infanticide, and between fairness and competition. In each case, it takes several steps to get from the one to the other, but wherever we turn, social institutions result from an interplay between opposing forces. Evolution is a dialectical process.

Human nature, too, is inherently multidimensional, and the same applies to chimpanzee and bonobo nature. Even if the chimp's nature is more violent and the bonobo's more peaceful, chimps do resolve conflicts and bonobos do compete. In fact, chimpanzee peacemaking is all the more impressive given their obvious violent streak. Both apes have both tendencies, but each reaches a different balance.

Being both more systematically brutal than chimps and more empathic than bonobos, we are by far the most bipolar ape. Our societies are never completely peaceful, never completely competitive, never ruled by sheer selfishness, and never perfectly moral. Pure states are not nature's way. What's true for human society is also true for human nature. One can find both kindness and cruelty, nobility and vulgarity—sometimes all in the same person. We're full of contradictions, but mostly tamed ones. Talk of "tamed contradictions" may sound obscure, even mystical, but they're all around us. The solar system is a perfect example. It results from two opposing forces, one directed inward, the other outward. The sun's gravity balances the planets' centrifugal motions so perfectly that the entire solar system has stayed together for billions of years.

On top of the inherent duality of human nature comes the role of intelligence. Even if we customarily overestimate our rationality, there is no denying that human behavior is a combination of drive and intelligence. We exert little control over ancient urges for power, sex, safety, and food, but we habitually weigh the pros and cons of our actions before we engage in them. Human behavior is seriously modified by experience. This may sound too obvious to even mention, but it is a radically different way of putting things than the way biologists used to talk. In the 1960s, almost every noticeable tendency of the human species was labeled an "instinct," and Konrad Lorenz's *Instinktlehre* (German for "instinct doctrine") even included a "parliament" of instincts to tie them all together. The problem with the term "instinct," however, is that it downplays the role of learning and experience. A similar trend exists in some contemporary circles, this time favoring the term "module." The human brain is compared to a Swiss Army knife to which evolution has one by one added modules for everything from face-recognition and tool-use to child care and friendship. Unfortunately, no one knows exactly what a brain module is, and the evidence for their existence is no more tangible than the evidence for instincts.

It is undeniable that we have inborn predispositions, yet I don't see us as blind actors carrying out nature's genetic programs. I see us rather as improvisers who flexibly adjust to other improvisers on the scene with our genes offering hints and suggestions. The same applies to our fellow primates. Let me explain this with the example of Yeroen—at the Arnhem Zoo—who had injured his hand in a fight. Yeroen was building a coalition with the up-and-coming Nikkie, but in the shuffles leading to their partnership, Nikkie had bitten him. Although it wasn't a deep wound, Yeroen limped heavily. After a couple of days, however, we got the impression that he limped mainly when Nikkie was around. I found this hard to believe, so we set out to conduct systematic observations. Each time we saw Yeroen walk with a limp, we recorded Nikkie's whereabouts. This revealed that Nikkie's field of vision mattered a great deal. Yeroen would walk past the sitting Nikkie, for example, from a point in front of him to a point behind him and the whole time he was in Nikkie's view he would hobble pitifully, but once he had passed Nikkie, he would walk perfectly normally again.

Yeroen seemed to be faking a limp so that his partner would go easy on him and perhaps show him some sympathy. Hurting one's buddy is never a smart move, and Yeroen seemed to be pointing this out to Nikkie by exaggerating the pain he was in. Putting on a façade is of course familiar to us, since we do it all the time. A couple tries to look happy in public to hide a strained marriage, or people laugh at an unfunny joke told by their boss. Keeping up appearances is something we share with the apes.

Recently, we looked at hundreds of records of wrestling matches among young chimpanzees to see when they laugh the most. Playing apes open their mouth in an expression that looks like the human laugh. We were particularly interested in youngsters far apart in age, as it's not unusual for their games to get too rough. As soon as this happens, the mother of the younger partner steps in, sometimes hitting its playmate over the head. Obviously, the older chimp wants to avoid this. We found that juveniles playing with infants laughed a great deal whenever the infant's mother was watching. It was as if they were saying "Look how much fun we're having!" They laughed far less if they found themselves alone with the infant. Their behavior thus depended on whether mom could see them. Under her eyes, they projected a jolly mood so as to be left alone.

Pretense during play or among political rivals is one reason I have trouble with the theory of animals as blind actors. Instead of being genetically programmed when to limp or when to laugh, apes are acutely aware of their social environment. Like humans, they ponder many options in front of them and decide what to do dependent on the circumstances. In the laboratory, apes are usually tested on abstract problems, such as finding rewards pointed out by experimenters or seeing the difference among four, five, or six items (a capacity known as "numerosity"). If they fail, as they sometimes do, the conclusion is often that we're smarter than they are. In the social domain, however, in which apes deal with those they've known all their life, they give the impression of being about as intelligent as we are.

One crude way to test this would be to place a human in a chimp colony. This is obviously unrealistic as a chimp's strength far outstrips a man's, but imagine that we could find someone strong enough to stand up to adult chimps. We might be able to see how well he or she would fare within an ape community. The challenge for the human would be to win over friends without being too submissive, because without some level of assertiveness, one will just end up at the bottom of the pecking order or worse. Success will require that, just as in real life, the person be neither a bully nor a doormat. There would be no point trying to hide fear or hostility, because human body language is an open book to chimps. My prediction is that an ape colony would prove no easier to master than an average collection of people at work or at school.

All of this is to say that ape social life is full of intelligent decision-making. For this reason, comparisons among humans, chimps, and bonobos go well beyond shared "instincts" or "modules," however defined. All three species face similar social dilemmas and need to overcome similar contradictions while going after status, mates, and resources. They apply their full brainpower to find solutions. True, our species looks farther ahead and weighs more options than the apes do, but this hardly seems a fundamental difference. Even if we wield the better chess computer, we're still all playing chess.

## FOREVER YOUNG

Many people believe that our species is still evolving to ever greater heights, while the apes have simply stood still. But did our fellow primates really stop evolving? And is our species really still on the move? Ironically, the opposite may be true. Perhaps our evolution has stopped, whereas evolutionary pressures still operate on the apes.

Evolution works through the survival of variants that outreproduce other variants. A couple of centuries ago this still applied to our species. In unhealthy places, such as rapidly growing urban areas, the death rate of people outstripped their birthrate. This meant that some had larger families than others, and yet others no families at all. Nowadays, in contrast, the number of children reaching the age of twenty-five is close to 98 percent. Under such circumstances, everyone has a shot at being represented in the gene pool.

Good nutrition and modern medicine have removed the selection pressures driving human evolution. For example, women and babies used to run serious risks during labor. A narrow birth canal (relative to our outsized skull) was partly to blame, meaning that there was continuous evolutionary pressure to keep the canal wide. Cesarean sections changed all that. In the United States, 26 percent of births are by C-section, and in some private clinics in Brazil the rate is as high as 90 percent. More and more women with narrow birth canals will survive, passing on a trait that a few generations ago was a death warrant. The inevitable result will be a growing number of C-sections until natural birth becomes the exception.

Continued evolution requires a *danse macabre* around those who died before they reproduced. This may still happen in the industrialized world—for example, in the form of a devastating flu pandemic. Those with superior immunity will survive to pass on their genes, such as happened during the Black Death of the fourteenth century, which in a span of five years killed an estimated 25 million people in Europe alone. Some scientists believe that a highly infectious Ebolalike virus was transmitted from person to person. Immunity to such viruses is presently more common in Europe than in other parts of the world, possibly owing to this massive moment of natural selection.

Similarly, we can expect growing resistance to the human immunodeficiency virus (HIV) in sub-Saharan Africa, where nearly 10 percent of the population is infected. It is known that a small minority of people resist infection and that another minority fails to develop AIDS even if infected. Biologists label such people "adaptive mutations." They will propagate until their descendants cover the continent. The process will be completed only after enormous loss of life, however. Wild chimpanzees in Africa must already have gone through this: they carry the closely related simian immunodeficiency virus (SIV) without any ill effects.

Apart from immunity, which will likely keep adjusting, it is unclear which genetic changes we can still expect in our species. Humanity may have reached its biological peak, so to speak, one we will not be able to surpass unless we set up deliberate breeding programs (which I hope we'll never do). Despite funny books such as *The Darwin Awards,* which describe people who remove themselves from the gene pool by engaging in incomprehensibly stupid acts (such as the shoplifter who ran off while stuffing a pair of live lobsters with good-sized claws into his pants, thus causing an unplanned vasectomy), a few such accidents won't improve the human race. So long as there's no connection between intelligence and the number of children people bear, human brain size will remain what it is today.

But what about culture? When cultural change was still slow, human biology kept up. Some cultural and genetic traits were transmitted together, a phenomenon known as "dual inheritance." Our ancestors became lactose resistant, for example, when they started raising cattle. Every young mammal is able to digest milk, but the necessary enzyme typically turns off after weaning. In humans, this occurs after the age of four. Those who can't handle lactose suffer diarrhea and vomiting whenever they drink fresh cow's milk. This is the original condition of our species, and it's typical of most adults in the world. Only the descendants of cattle herders, such as Northern Europeans and dairy-dependent African pastoralists, are able to absorb vitamin D and calcium from milk, a genetic change going back ten thousand years to when sheep and cattle were first domesticated.

Today's cultural developments, however, are too fast for biology to keep pace. Text-messaging via cell phones is unlikely to make our thumbs grow longer. Rather, we developed text-messaging for the thumbs our species already possesses. We have become experts at changing the environment to our advantage. Therefore, I don't believe in the continued evolution of the human race—certainly not as it would affect body shape and behavior. We have removed the only handle biology has to modify us, which is differential reproduction.

It is unclear if ape evolution will keep going even though apes are still subject to real pressures. The problem is that they are under too much pressure, giving them hardly any chance: they hover on the brink of extinction. For years, I have clung to the idea that given the large remaining stretches of rain forest in the world, there would always be apes to accompany us. But I have become pessimistic. Because of massive habitat destruction, large fires, illegal poaching, the bush meat trade (people in Africa actually *eat* apes), and most recently the Ebola virus sweeping through ape populations, there may be only two hundred thousand chimpanzees left in the wild, one hundred thousand gorillas, twenty thousand bonobos, and an equal number of orangutans. If this sounds like a lot of apes, compare this with the enemy—humans—which number more than six billion. It is an unequal battle, and the prediction is that by the year 2040 virtually every suitable ape habitat will be gone.

We humans will be diminished if we cannot even protect the animals closest to us, who share almost all of our genes, and who differ from us only by degree. If we let them go, we can just as well let everything go and turn the idea that we are the only intelligent life on earth into a self-fulfilling prophecy. Even though I have studied apes in captivity all my life, I've seen enough of them in nature to feel that their life there—their dignity, their belonging, their role—is irreplaceable. To lose this would be to lose a big chunk of ourselves.

Wild populations of apes are priceless when it comes to illuminating past evolution. We barely know, for example, why bonobos and chimpanzees differ so much from each other. What happened two million years ago when their branches split? Was the original ape more like a chimp or more like a bonobo? We know that bonobos presently live in a richer habitat than chimps, one that allows mixed groups of males and females to forage together. This permits greater social cohesion than in chimps, who in their quest for food split up into small parties. The "sisterhood" among unrelated females that is typical of bonobo society would not have been possible without predictable, abundant food sources. Bonobos have access to enormous fruit trees that allow many individuals to feed together, and they also consume herbs that are plentiful on the forest floor. Since the same herbs are also a staple of the gorilla's diet, it has been speculated that the total absence of gorillas in the habitat of bonobos has left the bonobos a niche that remained closed to chimps, who compete with gorillas throughout their range.

Bonobos have another interesting feature that connects them to us, which is that they are a "forever young" primate. This is known as the "neoteny" argument—applied to our own species ever since a Dutch anatomist made the shocking claim in 1926 that Homo sapiens looks like a primate fetus that has reached sexual maturity. Stephen Jay Gould considered retention of juvenile traits the hallmark of human evolution. He didn't know about bonobos, the adults of which retain the small, rounded skull of a young chimp as well as the white tail tufts that chimps lose after the age of five. The voices of adult bonobos are as high-pitched as those of infant chimps, they stay playful all their lives, and even the females' frontally oriented vulva is considered neotenous, a trait also present in our own species.

Human neoteny is reflected in our naked skin and especially in our ballooning cranium and flat face. Human adults look like very young apes. Is the crown of creation arrested in its development? There is no doubt that our success as a species is tied to the fact that we have carried the inventiveness and curiosity of young mammals forward into adulthood. We have been named *Homo ludens:* the playful ape. We play games until we die, we dance and sing, and we add to our knowledge by reading nonfiction or taking senior university classes.

There's a great need for us to stay young at heart. Given that humanity cannot pin its hopes on continued biological evolution, it needs to build upon its existing primate heritage. Being only loosely programmed and having sipped from evolution's youth potion, this heritage is rich and varied and full of flexibility.

## A WHIFF OF IDEOLOGY

Owing to their near perfect coordination and willingness to sacrifice themselves for the whole, ant colonies are often compared to socialist societies. Both are a workers' paradise. Yet next to the order of an anthill, even the best drilled human workforce looks like inefficient anarchy. People go home after work, drink, gossip, get lazy—none of which any self-respecting ant would ever do. Notwithstanding the massive indoctrination efforts by Communist regimes, people refuse to submerge themselves for the greater good. We are sensitive to collective interests, but not to the point of giving up our individual ones. Communism went under due to an economic incentive structure that was out of touch with human nature. Unfortunately, it did so only after causing immense suffering and death.

Nazi Germany was quite a different ideological disaster. Here, too, the collective (*das Volk*) was placed above the individual, but instead of relying on social engineering, the methods of choice were scapegoating and genetic manipulation. People were divided into "superior" and "inferior" types, the first of which needed to be protected against contamination by the second. In the horrible medical language of the Nazis, a healthy *Volk* required removal of cancerous elements. This idea was followed to its extreme in a manner that has given biology a nasty reputation in Western societies.

Don't think that the underlying selectionist ideology was restricted to this particular time and place, however. In the early part of the twentieth century, the eugenics movement—which sought to improve humanity by "breeding from the fitter stocks"—enjoyed widespread appeal in both the United States and Great Britain. Based on ideas going back to Plato's

*Republic,* castration of criminals was considered acceptable. And Social Darwinism—the idea that in a laissez-faire economy the strong will outcompete the weak, resulting in general improvement of the population—still inspires political agendas today. In this view, the poor should not be aided so as not to upset the natural order.

Political ideology and biology are awkward bedfellows, and most biologists prefer to sleep in a separate room. The reason we have been unsuccessful at this is the incredible appeal of the words "nature" and "natural." They sound so reassuring that every ideology wants to embrace them. This means that biologists who write about behavior and society risk being sucked into the political spin zone. This happened, for example, after publication of our study on simian fairness. Having demonstrated that a monkey will refuse cucumber as soon as its neighbor gets grapes, newspapers used our findings to call for a more egalitarian society. "If monkeys hate unfair treatment, why shouldn't we?" the op-ed pieces asked. This triggered strange reactions, such as one e-mail accusing us of being Communists. The writer felt that we were trying to undermine capitalism, which apparently doesn't care for fairness. What the critic had failed to notice, though, was that the reactions of our monkeys paralleled the workings of the free market. What could be more capitalist than comparing what you get with what someone else is getting and complaining if the price isn't right?

In 1879, American economist Francis Walker tried to explain why members of his profession were in such "bad odor amongst real people." He blamed it on their inability to understand why human behavior fails to comply with economic theory. We do not always act the way economists think we should, mainly because we're both less selfish and less rational than economists think we are. Economists are being indoctrinated into a cardboard version of human nature, which they hold true to such a degree that their own behavior has begun to resemble it. Psychological tests have shown that economics majors are more egoistic than the average college student. Exposure in class after class to the capitalist self-interest model apparently kills off whatever prosocial tendencies these students have to begin with. They give up trusting others, and conversely others give up trusting them. Hence the bad odor.

Social mammals, in contrast, know trust, loyalty, and solidarity. Like the chimps in Taï National Park, they don't leave the unfortunate behind. Moreover, they have ways of handling potential profiteers, such as withholding cooperation from those who fail to cooperate themselves. Reciprocity allows them to build the sort of support system that many economists see as a chimera. In the group life of our close relatives, it's not hard to recognize both the competitive spirit of capitalism and a well-developed community spirit. The political system that seems to fit us best would therefore have to balance the two. Not being ants, pure socialism doesn't suit us. Recent history has shown what happens when individual ambition is stifled. But even though the fall of the Berlin Wall was hailed as a triumph of the free market, there is no guarantee that pure capitalism will fare any better than socialism did.

Not that this form of government exists anywhere. Even the United States has an abundance of checks, balances, unions, and subsidies that constrain its marketplace. Compared to the rest of the world, though, the United States is an experiment in unbridled competition. This experiment has made it the richest nation in the history of civilization. Something puzzling is going on, however, as the nation's health is increasingly lagging behind its wealth.

The United States used to have the world's healthiest and tallest citizens, but now ranks at the bottom among industrialized nations in terms of longevity and height and at the top in terms of teenage pregnancy and infant mortality. Whereas most nations have been adding nearly one inch per decade to their average height, the United States has not done so since the 1970s. The result is that Northern Europeans now tower on average three inches over Americans. This isn't explained by recent United States immigrants, which constitute too tiny a fraction of the population to affect these statistics. In terms of life expectancy, too, the United States is not keeping up with the rest of the world. On this critical health index, Americans don't even rank among the top twenty-five nations anymore.

How to explain this? The first culprit that comes to mind is the privatization of health care, resulting in millions of uninsured people. But the problem may go deeper. British economist Richard Wilkinson, who has gathered global data on the connection between socioeconomic status and health, blames inequality. With its giant underclass, the income gap in the United States resembles that of many third-world nations. The top 1 percent of Americans has more income to spend than the bottom 40 percent taken together. This is a huge gap compared with Europe and Japan. Wilkinson argues that large income disparities erode the social fabric. They induce resentment and undermine trust, which causes stress to both the rich and the poor. No one feels at ease within such a system. The result is that the world's richest nation now has one of its poorest health records.

Whatever one thinks of a political system, if it fails to promote its citizens' physical well-being, it has a problem. And so in the same way that Communism collapsed due to a mismatch between ideology and human behavior, unmitigated capitalism may be unsustainable as it celebrates the material well-being of a few while shortchanging the rest. It denies the basic solidarity that makes life bearable. In doing so, it goes against a long evolutionary history of egalitarianism, which in turn relates to our cooperative nature. Primate experiments show how cooperation breaks down if benefits aren't shared among all participants, and human behavior likely follows the same principle.

The book of nature thus offers pages that will please both liberals and conservatives, both those who believe that we are all in the same boat and those who put faith in the pursuit of self-interest. When Margaret Thatcher dismissed society as a mere illusion, she was clearly not describing the intensely social primates that we are. And when Petr Kropotkin, a nineteenth-century

Russian prince, felt that the struggle for life could lead only to more and more cooperation, he was closing his eyes to free competition and its stimulating effects. The challenge is to find the right balance between the two.

Our societies probably work best if they mimic as closely as possible the small-scale communities of our ancestors. We certainly did not evolve to live in cities with millions of people where we bump into strangers everywhere we go, are threatened by them in dark streets, sit next to them in the bus, and give them the finger in traffic jams. Like bonobos in their cohesive communities, our ancestors were surrounded by people they knew and dealt with every day. It's remarkable that our societies are as orderly, productive, and relatively secure as they are. But city planners can and must do a better job at approximating the community life of old, in which everyone knew every child's name and home address.

The term "social capital" refers to the public safety and sense of security derived from a predictable environment and dense social network. Older neighborhoods in cities like Chicago, New York, London, and Paris do produce such social capital, but only because they were designed for people to live, work, do their shopping, and go to school in. This way people get to know each other and begin to share values. A young woman walking home in the evening will be surrounded by so many residents with a stake in safe streets that she can feel protected. She is surrounded by an unspoken neighborhood watch. The modern trend to physically separate places where human needs are satisfied disrupts this tradition, making us live at one place, shop at another, and work at yet another. It's a disaster for community building, not to mention the time, stress, and fuel it takes to move all those people around.

In the words of Edward Wilson, biology holds us "on a leash" and will let us stray only so far from who we are. We can design our life any way we want, but whether we will thrive depends on how well the life fits human predispositions.

I encountered a vivid example of this while having afternoon tea with a young couple during a visit to an Israeli kibbutz in the 1990s. They had both been raised on nearby kibbutzim when children were still being separated from their parents to grow up with other children in the cooperative. The couple explained that this practice had now been abandoned and that parents were permitted to keep their children at home after school and at night. The change was a relief, they said, because having your children close "just feels right."

How obvious! The kibbutz had felt the leash's limit. I hesitate to predict what we humans can and cannot do, but the mother-child bond would seem sacrosanct as it goes to the core of mammalian biology. We face the same sorts of limits when deciding what kind of society to build and how to achieve global human rights. We are stuck with a human psychology shaped by millions of years of life in small communities so that we somehow need to structure the world around us in a way recognizable to this psychology. If we could manage to see people on other continents as part of us, drawing them into our circle of reciprocity and empathy, we would be building upon, rather than going against, our nature.

In 2004, the Israeli justice minister caused a political uproar by sympathizing with the enemy. Yosef Lapid questioned the plans of the Israeli army to demolish thousands of Palestinian homes in a zone along the Egyptian border. He had been touched by images on the evening news. "When I saw a picture on the TV of an old woman on all fours in the ruins of her home looking under some floor tiles for her medicines, I did think, 'What would I say if it were my grandmother?'" Lapid's grandmother was a Holocaust victim. The nation's hard-liners did not like to hear these sentiments, of course, and went out of their way to distance themselves from them. The incident goes to show how a simple emotion can widen the definition of one's group. Lapid had suddenly realized that Palestinians were part of his circle of concern, too. Empathy is the one weapon in the human repertoire able to rid us of the curse of xenophobia.

Empathy is fragile, though. In our close relatives it is switched on by events within their community, such as a youngster in distress, but it is just as easily switched off with regard to outsiders or members of other species, such as prey. The way a chimpanzee bashes in the skull of a live monkey to get at its brain by hitting it against a tree trunk is no advertisement for ape empathy. Bonobos are less brutal, but in their case as well empathy needs to pass several filters before it will be expressed. Often, the filters stop it, because no ape can afford to feel pity for all living things all the time. This applies equally to humans. Our evolutionary design makes it hard to identify with outsiders. We've been designed to hate our enemies, to ignore the needs of people we barely know, and to distrust anybody who doesn't look like us. Even if within our communities we are largely cooperative, we become almost a different animal in our treatment of strangers.

Such attitudes were summarized by Winston Churchill, long before he earned his reputation as a brave warrior-politician, when he wrote, "The story of the human race is War. Except for brief and precarious interludes, there has never been peace in the world; and before history began, murderous strife was universal and unending." As we have seen, this is a gross exaggeration. No one denies our warrior potential, but Churchill definitely got the interludes wrong. Contemporary hunter-gatherer groups coexist in peace most of the time. This probably applied even more to our ancestors, who lived on a planet with lots of space and relatively little need for competition. They must have experienced long stretches of harmony between groups, interrupted only by brief interludes of confrontation.

Even though circumstances have changed—making the maintenance of peace much harder than in those days—a return to the mind-set that enabled mutually beneficial intergroup relations is perhaps less of a stretch than it may seem to those stressing our warrior side. After all, we have at least as long a history of getting along with other groups as of fighting them. We have both the chimpanzee side in us, which precludes friendly relations between groups, and the bonobo side, which permits sexual mingling and grooming across the border.

## WHICH INNER APE?

When Helena Bonham Carter was asked during an interview how she had prepared for her role of Ari in *Planet of the Apes,* she said that she had simply gotten in touch with her inner ape. She and the other actors went to a so-called simian academy to acquire ape postures and movements, but even though the petite Bonham Carter played a chimp, what she'd found within herself instead was, I believe, a sensual bonobo.

The contrast between those two apes reminds me of a distinction by psychologists between HE and HA personalities. HE stands for "Hierarchy-Enhancing," meaning a personality who believes in law and order and harsh measures to keep everybody in place. HA, on the other hand, stands for "Hierarchy-Attenuating," meaning a personality who seeks to level the playing field. The point is not which tendency is more desirable, because it's only together that they create human society as we know it. Our societies balance both types, having institutions that are either more HE, such as the criminal justice system, or more HA, such as civil rights movements and organizations that care for the poor.

Each person leans toward one type or the other, and we might even classify entire species this way, with chimps being more HE and bonobos more HA. Do we perhaps act like a hybrid between those two apes? We know little about the behavior of actual hybrids, but they are biologically possible and do in fact exist. No self-respecting zoo would intentionally crossbreed two endangered primates, but there's one report of a small French traveling circus featuring apes with curious voices. These apes were thought to be chimps, but to the expert ear their calls are as shrill as those of bonobos. It turns out that, unbeknownst to it, the circus long ago acquired a bonobo male, who became known as Congo. The trainer soon noticed this male's insatiable sex drive, which he exploited by rewarding Congo's best performances by a romp with the circus' female apes, all chimpanzees. The resulting progeny—perhaps to be called "bonanzees" or "chimpobos"—walk upright with remarkable ease and strike everyone by their gentility and sensitivity.

Perhaps we share a lot with these hybrids. We have the fortune of having not one but two inner apes, which together allow us to construct an image of ourselves that is considerably more complex than what we have heard coming out of biology for the past twenty-five years. The view of us as purely selfish and mean, with an illusory morality, is up for revision. If we are essentially apes, as I would argue, or at least descended from apes, as every biologist would argue, we are born with a gamut of tendencies from the basest to the noblest. Far from being a figment of the imagination, our morality is a product of the same selection process that shaped our competitive and aggressive side.

That such a creature could have been produced through the elimination of unsuccessful genotypes is what lends the Darwinian view its power. If we avoid confusing this process with its products—the Beethoven error—we see one of the most internally conflicted animals ever to walk the earth. It is capable of unbelievable destruction of both its environment and its own kind, yet at the same time it possesses wells of empathy and love deeper than ever seen before. Since this animal has gained dominance over all others, it's all the more important that it takes an honest look in the mirror, so that it knows both the archenemy it faces and the ally that stands ready to help it build a better world.

# Sources

## One: Apes in the Family

Ardrey, R. (1961) *African Genesis: A Personal Investigation into the Animal Origins and Nature of Man.* New York: Simon & Schuster.

Baron-Cohen, S. (2003) *The Essential Difference: The Truth About the Male and Female Brain.* New York: Basic Books.

Cartmill, M. (1993) *A View to a Death in the Morning: Hunting and Nature Through History.* Cambridge, MA: Harvard University Press.

Cohen, S., Doyle, W. J., Skoner, D. P., Rabin, B. S., and Gwaltney, J. M. (1997) "Social Ties and Susceptibility to the Common Cold." *Journal of the American Medical Association* 277: 1940–1944.

Coolidge, H. J. (1933) "*Pan Paniscus:* Pygmy Chimpanzee from South of the Congo River." *American Journal of Physical Anthropology* 18: 1–57.

———. (1984) "Historical Remarks Bearing on the Discovery of *Pan Paniscus.*" In *The Pygmy Chimpanzee,* Susman, R. L. (Ed.), pp. ix–xiii. New York: Plenum.

Darwin, C. (1967 [1859]) *On the Origin of Species by Means of Natural Selection or the Preservation of Favoured Races in the Struggle for Life.* London: John Murray.

———. (1981 [1871]) *The Descent of Man, and Selection in Relation to Sex.* Princeton, NJ: Princeton University Press.

Dawkins, R. (1976) *The Selfish Gene.* Oxford: Oxford University Press.

de Waal, F. B. M. (1980) "Aap Geeft Aapje de Fles." *De Levende Natuur* 82(2): 45–53.

———. (1996) *Good Natured: The Origins of Right and Wrong in Humans and Other Animals.* Cambridge, MA: Harvard University Press.

———. (1997) *Bonobo: The Forgotten Ape,* with photographs by Frans Lanting. Berkeley, CA: University of California Press.

Ghiselin, M. (1974) *The Economy of Nature and the Evolution of Sex.* Berkeley, CA: University of California Press.

Goodall, J. (1979) "Life and Death at Gombe." *National Geographic* 155(5): 592–621.

———. (1986) *The Chimpanzees of Gombe: Patterns of Behavior.* Cambridge, MA: Harvard University Press.

———. (1999) *Reason for Hope.* New York: Warner.

Greene, J., and Haidt, J. (2002) "How (and Where) Does Moral Judgement Work?" *Trends in Cognitive Sciences* 16: 517–523.

Hoffman, M. L. (1978) "Sex Differences in Empathy and Related Behaviors." *Psychological Bulletin* 84: 712–722.

Kano, T. (1992) *The Last Ape: Pygmy Chimpanzee Behavior and Ecology.* Stanford, CA: Stanford University Press.

Köhler, W. (1959 [1925]) *Mentality of Apes.* 2nd edition. New York: Vintage.

Menzel, C. R. (1999) "Unprompted Recall and Reporting of Hidden Objects by a Chimpanzee (*Pan Troglodytes*) After Extended Delays." *Journal of Comparative Psychology* 113: 426–434.

Montagu, A., Editor. (1968) *Man and Aggression.* London: Oxford University Press.

Morris, D. (1967) *The Naked Ape.* New York: McGraw-Hill.

Nakamichi, M. (1998) "Stick Throwing by Gorillas at the San Diego Wild Animal Park." *Folia primatologica* 69: 291–295.

Nesse, R. M. (2001) "Natural Selection and the Capacity for Subjective Commitment." In *Evolution and the Capacity for Commitment,* Nesse, R. M. (Ed.), pp. 1–44. New York: Russell Sage.

Nishida, T. (1968) "The Social Group of Wild Chimpanzees in the Mahali Mountains." *Primates* 9: 167–224.

Parr, L. A., and de Waal, F. B. M. (1999) "Visual Kin Recognition in Chimpanzees." *Nature* 399: 647–648.

Patterson, T. (1979) "The Behavior of a Group of Captive Pygmy Chimpanzees (*Pan Paniscus*)." Primates *20: 341–354.*

Ridley, M. (1996) *The Origins of Virtue.* London: Viking.

———. (2002) *The Cooperative Gene.* New York: Free Press.

Schwab, K. (February 24, 2003) "Capitalism Must Develop More of a Conscience." *Newsweek.*

Smith, A. (1937 [1759]) *A Theory of Moral Sentiments.* New York: Modern Library.

Sober, E., and Wilson, D. S. (1998) *Unto Others: The Evolution and Psychology of Unselfish Behavior.* Cambridge, MA: Harvard University Press.

Taylor, S. (2002) *The Tending Instinct.* New York: Times Books.

Tratz, E. P., and Heck, H. (1954) "Der Afrikanische Anthropoide 'Bonobo,' eine Neue Menschenaffengattung." *Säugetierkundliche Mitteilungen* 2: 97–101.

Wildman, D. E., Uddin, M., Liu, G., Grossman, L. I., and Goodman, M. (2003) "Implications of Natural Selection in Shaping 99.4% Nonsynonymous DNA Identity Between Humans and Chimpanzees: Enlarging Genus Homo." *Proceedings of the National Academy of Sciences* 100: 7181–7188.

Williams, G. C. (1988) Reply to comments on "Huxley's Evolution and Ethics in Sociobiological Perspective." *Zygon* 23: 437–438.

Wilson, E. O. (1978) *On Human Nature.* Cambridge, MA: Harvard University Press.

Wrangham, R. W., and Peterson, D. (1996) *Demonic Males: Apes and the Evolution of Human Aggression.* Boston: Houghton Mifflin.

Wright, R. (1994) *The Moral Animal: The New Science of Evolutionary Psychology.* New York: Pantheon.

Yerkes, R. M. (1925) *Almost Human.* New York: Century.

Zihlman, A. L. (1984) "Body Build and Tissue Composition in *Pan Pansicus* and *Pan Troglodytes,* with Comparisons to Other Hominoids." In *The Pygmy Chimpanzee,* Susman, R. L. (Ed.), pp. 179–200. New York: Plenum.

Zihlman, A. L., Cronin, J. E., Cramer, D. L., and Sarich, V. M. (1978) "Pygmy Chimpanzee as a Possible Prototype for the Common Ancestor of Humans, Chimpanzees, and Gorillas." *Nature* 275: 744–746.

### Two: Power

Adang, O. (1999) *De Machtigste Chimpansee van Nederland: Leven en Dood in een Mensapengemeenschap.* Amsterdam: Nieuwezijds.

Boehm, C. (1993) "Egalitarian Behavior and Reverse Dominance Hierarchy." *Current Anthropology* 34: 227–254.

———. (1994) "Pacifying Interventions at Arnhem Zoo and Gombe." In *Chimpanzee Cultures,* Wrangham, R. W., McGrew, W. C., de Waal, F. B. M., and Heltne, P. (Eds.), pp. 211–226. Cambridge, MA: Harvard University Press.

———. (1999) *Hierarchy in the Forest: The Evolution of Egalitarian Behavior.* Cambridge, MA: Harvard University Press.

de Waal, F. B. M. (1984) "Sex-Differences in the Formation of Coalitions Among Chimpanzees." *Ethology & Sociobiology* 5: 239–255.

———. (1994) "The Chimpanzee's Adaptive Potential: A Comparison of Social Life Under Captive and Wild Conditions." In *Chimpanzee Cultures,* Wrangham, R. W., McGrew, W. C., de Waal, F. B. M., and Heltne, P. (Eds.), pp. 243–260. Cambridge, MA: Harvard University Press.

———. (1997) *Bonobo: The Forgotten Ape,* with photographs by Frans Lanting. Berkeley, CA: University of California Press.

———. (1998 [1982]) *Chimpanzee Politics: Power and Sex Among Apes.* Revised edition. Baltimore, MD: Johns Hopkins University Press.

de Waal, F. B. M. and L. M. Luttrell. (1988) "Mechanisms of Social Reciprocity in Three Primate Species: Symmetrical Relationship Characteristics or Cognition?" *Ethology & Sociobiology* 9: 101–118.

———. (1989) "Toward a Comparative Socioecology of the Genus *Macaca*: Different Dominance Styles in Rhesus and Stumptail Monkeys." *American Journal of Primatology* 19: 83–109.

Doran, D. M., Jungers, W. L., Sugiyama, Y., Fleagle, J. G., and Heesy, C. P. (2002) "Multivariate and Phylogenetic Approaches to Understanding Chimpanzee and Bonobo Behavioral Diversity." In *Behavioural Diversity in Chimpanzees and Bonobos,* Boesch, C., Hohmann, G., and Marchant, L. F. (Eds.), pp. 14–34. Cambridge: Cambridge University Press.

Dowd, M. (April 10, 2002) "The Baby Bust." *The New York Times.*

Furuichi, T. (1989) "Social Interactions and the Life History of Female *Pan Paniscus* in Wamba, Zaire." *International Journal of Primatology* 10: 173–197.

———. (1992) "Dominance Status of Wild Bonobos at Wamba, Zaire." XIVth Congress of the International Primatological Society, Strasbourg, France.

———. (1997) "Agonistic Interactions and Matrifocal Dominance Rank of Wild Bonobos at Wamba." *International Journal of Primatology* 18: 855–875.

Gamson, W. (1961) "A Theory of Coalition Formation." *American Sociological Review* 26: 373–382.

Goodall, J. (1992) "Unusual Violence in the Overthrow of an Alpha Male Chimpanzee at Gombe." In *Topics in Primatology, Volume 1, Human Origins,* Nishida, T., McGrew, W. C., Marler, P., Pickford, M., and de Waal, F. B. M. (Eds.), pp. 131–142. University of Tokyo Press, Tokyo.

Grady, M. F., and McGuire, M. T. (1999) "The Nature of Constitutions." *Journal of Bioeconomics* 1: 227–240.

Gregory, S. W., and Webster, S. (1996) "A Nonverbal Signal in Voices of Interview Partners Effectively Predicts Communication Accommodation and Social Status Perceptions." *Journal of Personality and Social Psycholology* 70: 1231–1240.

Gregory, S. W., and Gallagher, T. J. (2002) "Spectral Analysis of Candidates' Nonverbal Vocal Communication: Predicting U. S. Presidential Election Outcomes." *Social Psychology Quarterly* 65: 298–308.

Hobbes, T. (1991 [1651]) *Leviathan.* Cambridge: Cambridge University Press.

Hohmann, G., and Fruth, B. (1996) "Food Sharing and Status in Unprovisioned Bonobos." In *Food and the Status Quest,* Wiessner, P., and Schiefenhövel, W. (Eds.), pp. 47–67. Providence, RI: Berghahn.

Kevin, a young adult male (bonobo, San Diego)

An upright female (left) and male show long-legged, humanlike body proportions (bonobos, San Diego)

Yeroen, the old fox (chimpanzee, Arnhem)

Mama with daughter Moniek (chimpanzees, Arnhem)

Nikkie (left) bluffs at Luit, who pantgrunts in submission (chimpanzees, Arnhem)

Nikkie (center) grooms Yeroen, who has helped him rise above Luit (right) (chimpanzees, Arnhem)

Grooming is the social cement of any primate society—here between mother and daughter (chimpanzees, Yerkes)

A sharing cluster around favored branches and leaves (chimpanzees, Yerkes)

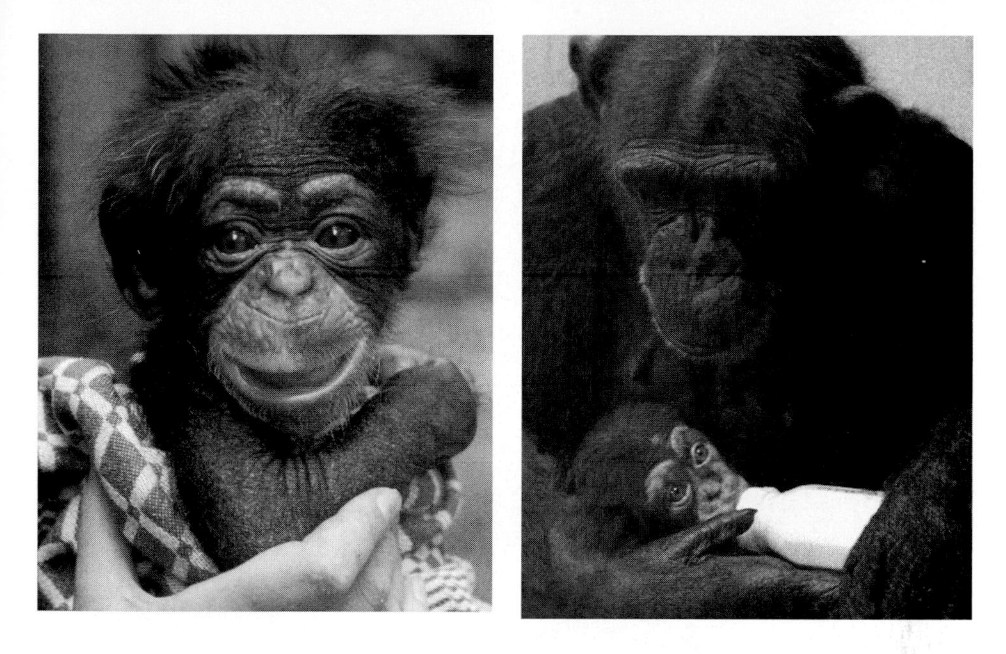

Roosje was bottle-raised by Kuif (chimpanzees, Arnhem)

Two genito-genital-rubbing females (bonobos, San Diego)

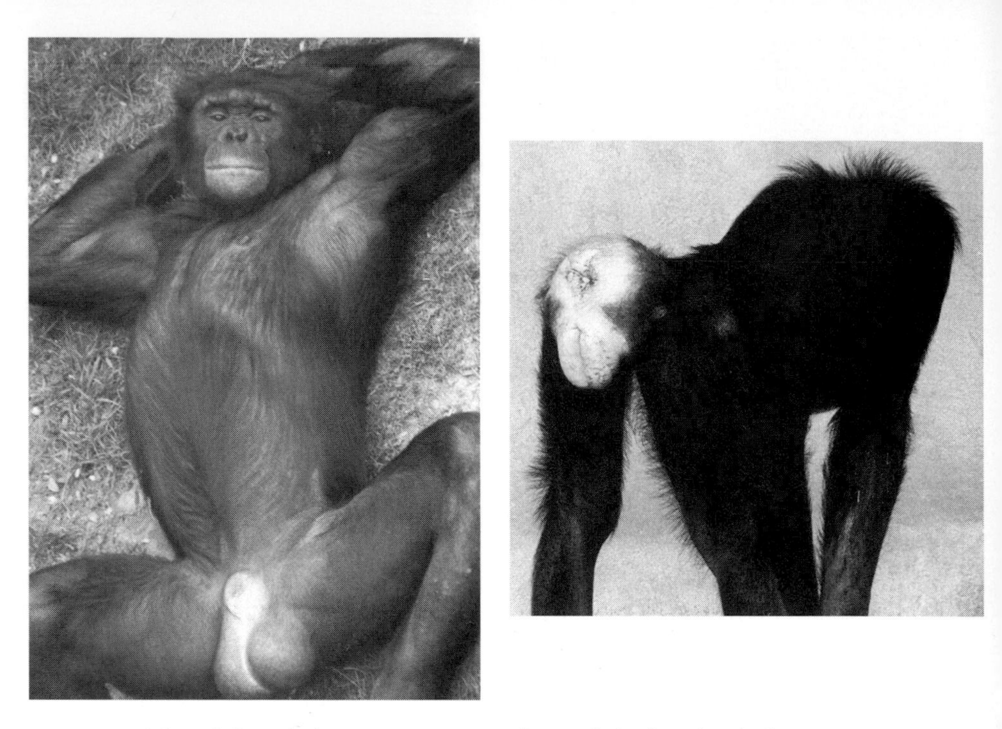

The adult male has impressive testicles, and the female a ballooning genital swelling (bonobos, San Diego)

Heterosexual sex in the "missionary" position (bonobos, San Diego)

Kano, T. (1996) "Male Rank Order and Copulation Rate in a Unit-Group of Bonobos at Wamba, Zaïre." In *Great Ape Societies,* McGrew, W. C., Marchant, L. F., and Nishida, T. (Eds.), pp. 135–145. Cambridge: Cambridge University Press.

Kano, T. (1998) Comments on C. B. Stanford. *Current Anthropology* 39: 410–411.

Kawanaka, K. (1984) "Association, Ranging, and the Social Unit in Chimpanzees of the Mahale Mountains, Tanzania." *International Journal of Primatology* 5: 411–434.

Konner, M. (2002) "Some Obstacles to Altruism." In *Altruistic Love: Science, Philosophy, and Religion in Dialogue,* Post, S. G., et al. (Eds.), pp 192–211. Oxford: Oxford University Press.

Lee, P. C. (1997) "The Meanings of Weaning: Growth, Lactation and Life History." *Evolutionary Anthropology* 5: 87–96.

Lee, R. B. (1979) *The !Kung San: Men, Women, and Work in a Foraging Society.* Cambridge: Cambridge University Press.

Mulder, M. (1979) *Omgaan met Macht.* Amsterdam: Elsevier.

Nishida, T. (1983) "Alpha Status and Agonistic Alliances in Wild Chimpanzees." *Primates* 24: 318–336.

Nishida, T., and Hosaka, K. (1996) "Coalition Strategies Among Adult Male Chimpanzees of the Mahale Mountains, Tanzania." In *Great Ape Societies,* McGrew, W. C., Marchant, L. F., and Nishida, T. (Eds.), pp. 114–134. Cambridge: Cambridge University Press.

Parish, A. R. (1993) "Sex and Food Control in the 'Uncommon Chimpanzee': How Bonobo Females Overcome a Phylogenetic Legacy of Male Dominance." *Ethology & Sociobiology* 15: 157–179.

Parish, A. R., and de Waal, F. B. M. (2000) "The Other 'Closest Living Relative': How Bonobos Challenge Traditional Assumptions About Females, Dominance, Intra- and Inter-Sexual Interactions, and Hominid Evolution." In *Evolutionary Perspectives on Human Reproductive Behavior,* LeCroy, D., and Moller, P. (Eds.), pp. 97–103. *Annals of the New York Academy of Sciences* 907.

Riss, D., and Goodall, J. (1977) "The Recent Rise to the Alpha-Rank in a Population of Free-Ranging Chimpanzees." *Folia primatologica* 27: 134–151.

Roy, R., and Benenson, J. F. (2002) "Sex and Contextual Effects on Children's Use of Interference Competition." *Developmental Psychology* 38: 306–312.

Sacks, O. (1985) *The Man who Mistook His Wife for a Hat.* London: Picador.

Sapolsky, R. M. (1994) *Why Zebras Don't Get Ulcers.* New York: Freeman.

Schama, S. (1987) *The Embarrassment of Riches: An Interpretation of Dutch Culture in the Golden Age.* New York: Knopf.

Schjelderup-Ebbe, T. (1922) "Beiträge zur Sozialpsychologie des Haushuhns." *Zeitschrift für Psychologie* 88: 225–252.

Sherif, M. (1966) *In Common Predicament: Social Psychology of Intergroup Conflict and Cooperation.* Boston: Houghton Mifflin.

Stanford, C. B. (1998) "The Social Behavior of Chimpanzees and Bonobos." *Current Anthropology* 39: 399–407.

Strier, K. B. (1992) "Causes and Consequences of Nonaggression in the Woolly Spider Monkey, or Muriqui." In *Aggression and Peacefulness in Humans and Other Primates,* Silverberg, J., and Gray, J. P. (Eds.), pp. 100–116. New York: Oxford University Press.

Thierry, B. (1986) "A Comparative Study of Aggression and Response to Aggression in Three Species of Macaque." In *Primate Ontogeny, Cognition and Social Behavior,* Else, J. G., and Lee, P. C. (Eds.), pp. 307–313. Cambridge: Cambridge University Press.

van Elsacker, L., Vervaecke, H., and Verheyen, R. F. (1995) "A Review of Terminology on Aggregation Patterns in Bonobos." *International Journal of Primatology* 16: 37–52.

Vervaecke, H., de Vries, H., and van Elsacker, L. (2000) "Dominance and Its Behavioral Measures in a Captive Group of Bonobos." *International Journal of Primatology* 21: 47–68.

Wiessner, P. (1996) "Leveling the Hunter: Constraints on the Status Quest in Foraging Societies." In *Food and the Status Quest,* Wiessner, P., and Schiefenhövel, W. (Eds.), pp. 171–191. Providence, RI: Berghahn.

Woodward, R., and Bernstein, C. (1976) *The Final Days.* New York: Simon & Schuster.

Zinnes, D. A. (1967) "An Analytical Study of the Balance of Power Theories." *Journal of Peace Research* 4: 270–288.

## Three: Sex

Alcock, J. (2001) *The Triumph of Sociobiology.* Oxford: Oxford University Press.

Alexander, M. G., and Fisher, T. D. (2003) "Truth and Consequences: Using the Bogus Pipeline to Examine Sex Differences in Self-Reported Sexuality." *Journal of Sex Research* 40: 27–35.

Angier, N. (1999) *Woman: An Intimate Geography.* New York: Houghton Mifflin.

Antilla, S. (2003) *Tales from the Boom-Boom Room: Women vs. Wall Street.* Princeton, NJ: Bloomberg Press.

Arribas, A. (2003) *Petite Histoire du Baiser.* Paris: Nicolas Philippe.

Bagemihl, B. (1999) *Biological Exuberance: Animal Homosexuality and Natural Diversity.* New York: St. Martin's Press.

Beckerman, S., and Valentine, P. (2002) *Cultures of Multiple Fathers: The Theory and Practice of Partible Paternity in Lowland South America.* Gainesville, FL: University Press of Florida.

Bereczkei, T., Gyuris, T., and Weisfeld, G. E. (2004) "Sexual Imprinting in Human Mate Choice." *Proceedings of the Royal Society of London* 271: 1129–1134.

Betzig, L. (1986) *Despotism and Differential Reproduction: A Darwinian View of History.* New York: Aldine de Gruyter.

Boesch, C., and Boesch, H. (1984) "Sex Differences in the Use of Natural Hammers by Wild Chimpanzees: A Preliminary Report." *Journal of Human Evolution* 13: 415–585.

Bray, O. E., Kennelly, J. J., and Guarino, J. L. (1975) "Fertility of Eggs Produced on Territories of Vasectomized Red-Winged Blackbirds." *Wilson Bulletin* 87: 187–195.

Brown Travis, C., Editor. (2003) *Evolution, Gender, and Rape.* Cambridge, MA: MIT Press.

Buss, D. M. (1989) "Sex Differences in Human Mate Preferences." *Behavioral and Brain Sciences* 12: 1–49.

Cardoso, F. L., and Werner, D. (2004) "Homosexuality." In *Encyclopedia of Sex and Gender: Men and Women in the World's Cultures*, Ember, C. R., and Ember, M. (Eds.), pp. 204–215. New York: Kluwer.

Dahl, J. F. (1986) "Cyclic Perineal Swelling During the Intermenstrual Intervals of Captive Female Pygmy Chimpanzees." *Journal of Human Evolution* 15: 369–385.

Dahl, J. F., Nadler, R. D., and Collins, D. C. (1991) "Monitoring the Ovarian Cycles of *Pan Troglodytes* and *Pan Paniscus*: A Comparative Approach." *American Journal of Primatology* 24: 195–209.

Daly, M., and Wilson, M. (1982) "Whom Are Newborn Babies Said to Resemble?" *Ethology & Sociobiology* 3: 69–78.

———. (1988) *Homicide.* Hawthorne, NY: Aldine de Gruyter.

de Waal, F. B. M. (1987) "Tension Regulation and Nonreproductive Functions of Sex Among Captive Bonobos." *National Geographic Research* 3: 318–335.

———. (1995) "Sex as an Alternative to Aggression in the Bonobo." In *Sexual Nature, Sexual Culture,* Abramson, P., and Pinkerton, S. (Eds.), pp. 37–56. Chicago: University of Chicago Press.

———. (1998) Commentary on C. B. Stanford. *Current Anthropology* 39: 407–408.

———. (April 2, 2000) "Survival of the Rapist," review of *A Natural History of Rape* by R. Thornhill and C. T. Palmer. *New York Times Book Review*, pp. 24–25.

———. (2001) *The Ape and the Sushi Master.* New York: Basic Books.

Deer, B. (March 9, 1997) "Death of the Killer Ape." *The Sunday Times Magazine* (London), March 9, 1997.

Diamond, M. (1990) "Selected Cross-Generational Sexual Behavior in Traditional Hawai'i: A Sexological Ethnography." In *Pedophilia: Biosocial Dimensions,* Feierman, J. R. (Ed.), pp. 378–393. New York: Springer.

Ehrlich, P. (2000) *Human Natures: Genes, Cultures, and the Human Prospect.* Washington, D.C.: Island Press.

Fisher, H. (1983) *The Sex Contract: The Evolution of Human Behavior.* New York: Quill.

Fossey, D. (1984) "Infanticide in Mountain Gorillas with Comparative Notes on Chimpanzees." In *Infanticide,* Hausfater, G., and Hrdy, S. B. (Eds.), pp. 217–235. New York: Aldine de Gruyter.

Foucault, M. (1978) *The History of Sexuality: An Introduction, Volume 1.* New York: Vintage.

Freese, J., and Meland, S. (2002) "Seven Tenths Incorrect: Heterogenity and Change in the Waist-to-Hip Ratios in *Playboy* Centerfold Models and Miss America Pageant Winners." *Journal of Sex Research* 39: 133–138.

Freud, S. (1950 [1913]) *Totem and Taboo: Some Points of Agreement Between the Mental Lives of Savages and Neurotics.* New York: Norton.

Friedman, D. M. (2001) *A Mind of its Own: A Cultural History of the Penis.* New York: Free Press.

Furuichi, T., and Hashimoto, C. (2002) "Why Female Bonobos Have a Lower Copulation Rate During Estrus Than Chimpanzees." In *Behavioural Diversity in Chimpanzees and Bonobos*, Boesch, C., Hohmann, G., and Marchant, L. F. (Eds.), pp. 156–167. Cambridge: Cambridge University Press.

Furuichi T., Idani, G., Ihobe, H., Kuroda, S., Kitamura, K., Mori, A., Enomoto, T., Okayasu, N., Hashimoto, C., and Kano, T. (1998) "Population Dynamics of Wild Bonobos at Wamba." *International Journal of Primatology* 19: 1029–1043.

Goldfoot, D. A., Westerborg-van Loon, H., Groeneveld, W., and Slob, A. K. (1980) "Behavioral and Physiological Evidence of Sexual Climax in the Female Stumptailed Macaque." *Science* 208: 1477–1479.

Gould, S. J. (1987) "Freudian Slip." *Natural History* April: 15–21.

Harcourt, A. H. (1995) "Sexual Selection and Sperm Competition in Primates: What Are Male Genitalia Good For?" *Evolutionary Anthropology* 4: 121–129.

Hashimoto, C., and Furuichi, T. (1994) "Social Role and Development of Noncopulatory Sexual Behavior of Wild Bonobos." In *Chimpanzee Cultures,* Wrangham, R. W., et al. (Eds.), pp. 155–168. Cambridge, MA: Harvard University Press.

Hawkes, K., O'Connell, J. F., Blurton-Jones, N. G., Alvarez, H., and Charnov, E. L. (1998) "Grandmothering, Menopause, and the Evolution of Human Life Histories." *Proceedings of the National Academy of Sciences* 95: 1336–1339.

Hobbes, T. (1991 [1651]) *Leviathan.* Cambridge: Cambridge University Press.

Hohmann, G., and Fruth, B. (2002) "Dynamics in Social Organization of Bonobos (*Pan Paniscus*)." In *Behavioural Diversity in Chimpanzees and Bonobos,* Boesch, C., Hohmann, G., and Marchant, L. F. (Eds.), pp. 138–150. Cambridge: Cambridge University Press.

Hrdy, S. B. (1979) "Infanticide Among Animals: A Review, Classification, and Examination of the Implications for the Reproductive Strategies of Females." *Ethology & Sociobiology* 1: 13–40.

———. (1999) *Mother Nature: A History of Mothers, Infants, and Natural Selection.* New York: Pantheon.

Hrdy, S. B., and Whitten, P. L. (1987) "Patterning of Sexual Activity." In *Primate Societies,* Smuts, B., et al. (Eds.), pp. 370–384. Chicago: University of Chicago Press.

Hua, C. (2001) *A Society Without Fathers or Husbands: The Na of China.* New York: Zone Books.

Jolly, A. (1999) *Lucy's Legacy: Sex and Intelligence in Human Evolution.* Cambridge, MA: Harvard University Press.

Kano, T. (1992) *The Last Ape: Pygmy Chimpanzee Behavior and Ecology.* Stanford, CA: Stanford University Press.

Kevles, B. (1986) *Females of the Species: Sex and Survival in the Animal Kingdom.* Cambridge, MA: Harvard University Press.

Kinsey, A. C., Pomeroy, W. B., and Martin, C. E. (1948) *Sexual Behavior and the Human Male.* Philadelphia: Saunders Company.

Kuroda, S. (1982) *The Unknown Ape: The Pygmy Chimpanzee.* (In Japanese) Tokyo: Chikuma-Shobo.

Laumann, E., Gagnon, J. H., Michael, R. T., and Michaels, S. (1994) *The Social Organization of Sexuality: Sexual Practices in the United States.* Chicago: University of Chicago Press.

Linden, E. (2002) *The Octopus and the Orangutan.* New York: Dutton.

Lovejoy, C. O. (1981) "The Origin of Man." *Science* 211: 341–350.

Malinowski, B. (1929) *The Sexual Life of Savages.* London: Lowe & Brydone.

Marlowe, F. (2001) "Male Contribution to Diet and Female Reproductive Success Among Foragers." *Current Anthropology* 42: 755–760

McGrew, W. C. (1979) "Evolutionary Implications of Sex-Differences in Chimpanzee Predation and Tool-Use." In *The Great Apes,* Hamburg, D. A., and McCown, E. R. (Eds.), pp. 440–463. Menlo Park, CA: Benjamin Cummings.

Michael, R. T., Gagnon, J. H., Laumann, B. O., and Kolata, G. (1994) *Sex in America: A Definitive Survey.* New York: Little, Brown.

Møller, A. P. (1988) "Ejaculate Quality, Testes Size and Sperm Competition in Primates." *Journal of Human Evolution* 17: 479–488.

Morris, D. (1967) *The Naked Ape.* New York: McGraw-Hill.

Nishida, T., and Kawanaka, K. (1985) "Within-Group Cannibalism by Adult Male Chimpanzees." *Primates* 26: 274–284.

Palombit, R. A. (1999) "Infanticide and the Evolution of Pair Bonds in Nonhuman Primates." *Evolutionary Anthropology* 7: 117–129.

Panksepp, J. (1998) *Affective Neuroscience: The Foundations of Human and Animal Emotions.* New York: Oxford University Press.

Potts, M., and Short, R. (1999) *Ever Since Adam and Eve: The Evolution of Human Sexuality.* Cambridge: Cambridge University Press.

Pusey, A. E., and Packer, C. (1994) "Infanticide in Lions: Consequences and Counter-Strategies." In *Infanticide and Parental Care,* Parmigiani, S., and vom Saal, F. (Eds.), pp. 277–299. Chur: Harwood Academic Publishers.

Reno, P. L., Meindl, R. S., McCollum, M. A., and Lovejoy, C. O. (2003) "Sexual Dimporphism in *Australopithecus Afarensis* Was Similar to That of Modern Humans." *Proceedings of the National Academy of Sciences* 100: 9404–9409.

Savage-Rumbaugh, S., and Wilkerson, B. (1978) "Socio-Sexual Behavior in *Pan Paniscus* and *Pan Troglodytes*: A Comparative Study." *Journal of Human Evolution* 7: 327–344.

Short, R. V. (1979) "Sexual Selection and its Component Parts, Somatic and Genital Selection as Illustrated by Man and the Great Apes." *Advances in the Study of Behaviour* 9: 131–158.

Simmons, L. W., Firman, R., Rhodes, G., and Peters. M. (2004) "Human Sperm Competition: Testis Size, Sperm Production and Rates of Extrapair Copulations." *Animal Behaviour* 68: 297–302.

Singh, D. (1993) "Adaptive Significance of Female Physical Attractiveness: Role of Waist-to-Hip Ratio." *Journal of Personality and Social Psychology* 65: 293–307.

Small, M. F. (1995) *What's Love Got to Do with It?* New York: Anchor Books.

———. (2003) "How Many Fathers are Best for a Child?" *Discover* April: 54–61.

Smuts, B. B. (1995) "The Evolutionary Origins of Patriarchy." *Human Nature* 6: 1–32.

Sommer, V. (1994) "Infanticide Among the Langurs of Jodhpur: Testing the Sexual Selection Hypothesis with a Long-Term Record." In *Infanticide and Parental Care,* Parmigiani, S., and vom Saal, F. S. (Eds.), pp. 155–187. Chur: Harwood Academic Publishers.

Stanford, C. B. (1999) *The Hunting Apes: Meat-eating and the Origins of Human Behavior.* Princeton, NJ: Princeton University Press.

Sugiyama, Y. (1967) "Social Organization of Hanuman Langurs." In *Social Communication Among Primates,* Altmann, S. A. (Ed.), pp. 221–253. Chicago: The University of Chicago Press.

Suzuki, A. (1971) "Carnivority and Cannibalism Observed Among Forest-Living Chimpanzees." *Journal of the Anthropological Society of Nippon* 79: 30–48.

Symons, D. (1979) *The Evolution of Human Sexuality.* New York: Oxford University Press.

Szalay, F. S., and Costello, R. K. (1991) "Evolution of Permanent Estrus Displays in Hominids." *Journal of Human Evolution* 20: 439–464.

Thompson-Handler, N. (1990) *The Pygmy Chimpanzee: Sociosexual Behavior, Reproductive Biology and Life History Patterns.* Unpublished dissertation, New Haven, CT: Yale University.

Thornhill, R., and Palmer, C. T. (2000) *The Natural History of Rape: Biological Bases of Sexual Coercion.* Cambridge, MA: MIT Press.

Tratz, E. P., and Heck, H. (1954) "Der Afrikanische Anthropoide 'Bonobo,' eine Neue Menschenaffengattung." *Säugetierkundliche Mitteilungen* 2: 97–101.

van Hooff, J. A. R. A. M. (2002) *De Mens, een Primaat Net Zo "Eigenaardig" als de Andere Primaten.* The Hague: Nederlandse Organisatie voor Wetenschappelijk Onderzoek (NWO).

van Schaik, C. P., and Dunbar, R. I. M. (1990) "The Evolution of Monogamy in Large Primates: A New Hypothesis and Some Crucial Tests." *Behaviour* 115: 30–62.

Vasey, P. L. (1995) "Homosexual Behavior in Primates: A Review of Evidence and Theory." *International Journal of Primatology* 16: 173–204.

Walker, A. (1998) *By the Light of My Father's Smile.* New York: Ballantine.

Wolf, A. P., and Durham, W. H. (2005) *Inbreeding, Incest, and the Incest Taboo.* Stanford, CA: Stanford University Press.

Wrangham, R. W. (1993) "The Evolution of Sexuality in Chimpanzees and Bonobos." *Human Nature* 4: 47–79.

Wright, C. (November 14–21, 2002) "Going Ape." www.bostonphoenix.com.

Yerkes, R. M. (1941) "Conjugal Contrasts Among Chimpanzees." *Journal of Abnormal and Social Psychology* 36: 175–199.

Zerjal, T., et al. (2003) "The Genetic Legacy of the Mongols." *American Journal of Human Genetics* 72: 717–721.

Zimmer, C. (2001) *Evolution: The Triumph of an Idea.* New York: Harper Collins.

Zuk, M. (2002) *Sexual Selections: What We Can and Can't Learn About Sex from Animals.* Berkeley, CA: University of California Press.

## Four: Violence

Atwood, M. E. (1989) *Cat's Eye.* New York: Doubleday.

Aureli, F. (1997) "Post-Conflict Anxiety in Nonhuman Primates: The Mediating Role of Emotion in Conflict Resolution." *Aggressive Behavior* 23: 315–328.

Aureli, F., and de Waal, F. B. M. (1997) "Inhibition of Social Behavior in Chimpanzees Under High-Density Conditions." *American Journal of Primatology* 41: 213–228.

———. (2000) *Natural Conflict Resolution.* Berkeley, CA: University of California Press.

Aureli, F., Preston, S. D., and de Waal, F. B. M. (1999) "Heart Rate Responses to Social Interactions in Free-Moving Rhesus Macaques (*Macaca Mulatta*): A Pilot Study." *Journal of Comparative Psychology* 113: 59–65.

Bauman, J. (1926) "Observations of the Strength of the Chimpanzee and its Implications." *Journal of Mammalogy* 7: 1–9.

Brewer, S. (1978) *The Forest Dwellers.* London: Collins.

Butovskaya, M., Verbeek, P., Ljungberg, T., and Lunardini, A. (2001) "A Multi-Cultural View of Peacemaking Among Young Children." In *Natural Conflict Resolution,* Aureli, F., and de Waal, F. B. M. (Eds.), pp. 243–258. Berkeley, CA: University of California Press.

Calhoun, J. B. (1962) "Population Density and Social Pathology." *Scientific American* 206: 139–148.

Cords, M., and Thurnheer, S. (1993) "Reconciliation with Valuable Partners by Long-Tailed Macaques." *Ethology* 93: 315–325.

de Waal, F. B. M. (1986) "Integration of Dominance and Social Bonding in Primates." *Quarterly Review of Biology* 61: 459–479.

———. (1986) "Prügelknaben bei Primaten und eine Tödliche Kampf in der Arnheimer Schimpansenkolonie." In *Ablehnung, Meidung, Ausschluß: Multidisziplinäre Untersuchungen über die Kehrseite der Vergemeinschaftung,* Gruter, M., and Rehbinder, M. (Eds.), pp.129–145. Berlin: Duncker & Humblot.

———. (1989) *Peacemaking Among Primates.* Cambridge, MA: Harvard University Press.

———. (1989) "The Myth of a Simple Relation Between Space and Aggression in Captive Primates." *Zoo Biology Supplement* 1: 141–148.

———. (1997) *Bonobo: The Forgotten Ape,* with photographs by Frans Lanting. Berkeley, CA: University of California Press.

———. (2000) "Primates—A Natural Heritage of Conflict Resolution." *Science* 289: 586–590.

———. (2001) *The Ape and the Sushi Master.* New York: Basic Books.

de Waal, F. B. M., and Johanowicz, D. L. (1993) "Modification of Reconciliation Behavior Through Social Experience: An Experiment with Two Macaque Species." *Child Development* 64: 897–908.

de Waal, F. B. M., and van Roosmalen, A. (1979) "Reconciliation and Consolation Among Chimpanzees." *Behavioral Ecology & Sociobiology* 5: 55–66.

Ember, C. R. (1978) "Myths About Hunter-Gathereres." *Ethnology* 27: 239–448.

Ferguson, B. R. (2002) "The History of War: Fact vs. Fiction." In *Must we Fight?* Ury, W. L. (Ed.), pp. 26–37. San Francisco, CA: Jossey-Bass.

———. (2003) "The Birth of War." *Natural History* July/August: 28–34.

Frye, D. P. (2001) "Conflict Management in Cross-Cultural Perspective." In *Natural Conflict Resolution,* Aureli, F., and de Waal, F. B. M. (Eds.), pp. 334–351. Berkeley, CA: University of California Press.

Gat, A. (1999) "The Pattern of Fighting in Simple, Small-Scale, Prestate Societies." *Journal of Anthropological Research* 55: 563–583.

Gavin, M. (2004) "Primate vs. Primate." *BBC Wildlife* January: 50–52.

Goodall, J. (1986) *The Chimpanzees of Gombe: Patterns of Behavior.* Cambridge, MA: Harvard University Press.

———. (1999) *Reason for Hope.* New York: Warner.

Haney, C., Banks, W. C., and Zimbardo, P. G. (1973) "Interpersonal Dynamics in a Simulated Prison." *International Journal of Criminology and Penology* 1: 69–97.

Hölldobler, B., and Wilson, E. O. (1994) *Journey to the Ants.* Cambridge, MA: Belknap Press.

Idani, G. (1990) "Relations Between Unit-Groups of Bonobos at Wamba: Encounters and Temporary Fusions." *African Study Monographs* 11: 153–186.

Johnson, R. (1972) *Aggression in Man and Animals.* Philadelphia, PA: Saunders Company.

Judge, P. G., and de Waal, F. B. M. (1993) "Conflict Avoidance Among Rhesus Monkeys: Coping with Short-Term Crowding." *Animal Behaviour* 46: 221–232.

Kamenya, S. (2002) "Human Baby Killed by Gombe Chimpanzee." *Pan Africa News* 9(2): 26.

Kano, T. (1992) *The Last Ape: Pygmy Chimpanzee Behavior and Ecology.* Stanford, CA: Stanford University Press.

Kayumbo, H. Y. (2002) "A Chimpanzee Attacks and Kills a Security Guard in Kigoma." *Pan Africa News* 9(2): 11–12.

Köhler, W. (1959 [1925]) *Mentality of Apes.* 2nd Edition, New York: Vintage.

Kutsukake, N., and Castles, D. L. (2004) "Reconciliation and Post-Conflict Third-Party Affiliation Among Wild Chimpanzees in the Mahale Mountains, Tanzania." *Primates* 45:157–165.

Lagerspetz, K. M., Björkqvist, K., and Peltonen, T. (1988) "Is Indirect Aggression Typical of Females?" *Aggressive Behavior* 14: 403–414.

Lever, J. (1976) "Sex Differences in the Games Children Play." *Social Problems* 23: 478–487.

Lorenz, K. Z. (1966 [1963]) *On Aggression.* London: Methuen.

Lux, K. (1990) *Adam Smith's Mistake.* Boston: Shambhala.

Maestripieri, D., Schino, G., Aureli, F., and Troisi, A. (1992) "A Modest Proposal: Displacement Activities as Indicators of Emotions in Primates." *Animal Behaviour* 44: 967–979.

Murphy, D. E., and Halbfinger, D. M. (June 16, 2002) "9/11 Aftermath Bridged Racial Divide, New Yorkers Say, Gingerly." *New York Times,* p. 21.

Nishida, T., Hiraiwa-Hasegawa, M., Hasegawa, T., and Takahata, Y. (1985) "Group Extinction and Female Transfer in Wild Chimpanzees in the Mahale Mountains National Park, Tanzania." *Zeitschrift für Tierpsychologie* 67: 274–285.

Palagi, E., Paoli, T., and Tarli, S. B. (2004) "Reconciliation and Consolation in Captive Bonobos (*Pan Paniscus*)." *American Journal of Primatology* 62: 15–30.

Power, M. (1991) *The Egalitarians: Human and Chimpanzee.* Cambridge: Cambridge University Press.

Preuschoft, S., Wang, X., Aureli, F., and de Waal, F. B. M. (2002) "Reconciliation in Captive Chimpanzees: A Reevaluation with Controlled Methods." *International Journal of Primatology* 23: 29–50.

Rabbie, J. M., and Horwitz, M. (1969) "The Arousal of Ingroup-Outgroup Bias by a Chance of Win or Loss." *Journal of Personality and Social Psychology* 69: 223–228.

Robarchek, C. A. (1979) "Conflict, Emotion, and Abreaction: Resolution of Conflict Among the Semai Senoi." *Ethos* 7: 104–123.

Rubin, L. B. (1985) *Just Friends.* New York: Harper and Row.

Sapolsky, R. M. (1993) "Endocrinology *Alfresco*: Psychoendocrine Studies of Wild Baboons." *Recent Progress in Hormone Research* 48: 437–462.

Sapolsky, R. M., and Share, L. J. (2004) "A Pacific Culture Among Wild Baboons: Its Emergence and Transmission." *Public Library of Science Biology* 2: 534–541.

Schneirla, T. C. (1944) "A Unique Case of Circular Milling in Ants, Considered in Relation to Trail Following and the General Problem of Orientation." *American Museum Novitates* 1253: 1–26.

Stanford, C. B. (1999) *The Hunting Apes: Meat-Eating and the Origins of Human Behavior.* Princeton, NJ: Princeton University Press.

Tannen, D. (1990) *You Just Don't Understand: Women and Men in Conversation.* New York: Ballantine.

Taylor, S. (2002) *The Tending Instinct.* New York: Times Books.

Verbeek, P., and de Waal, F. B. M. (2001) "Peacemaking Among Preschool Children." *Journal of Peace Psychology* 7: 5–28.

Verbeek, P., Hartup, W. W., and Collins, W. A. (2000) "Conflict Management in Children and Adolescents." In *Natural Conflict Resolution,* Aureli, F., and de Waal, F. B. M. (Eds.), pp. 34–53. Berkeley, CA: University of California Press.

Weaver, A., and de Waal, F. B. M. (2003) "The Mother-Offspring Relationship as a Template in Social Development: Reconciliation in Captive Brown Capuchins (*Cebus Apella*)." *Journal of Comparative Psychology* 117: 101–110.

Wilson, M. L., and Wrangham, R. W. (2003) "Intergroup Relations in Chimpanzees." *Annual Review of Anthropology* 32: 363–392.

Wittig, R. M., and Boesch, C. (2003) "'Decision-making' in Conflicts of Wild Chimpanzees (*Pan Troglodytes*): An Extension of the Relational Model." *Behavioral Ecology & Sociobiology* 54: 491–504.

Wrangham, R. W. (1999) "Evolution of Coalitionary Killing." *Yearbook of Physical Anthropology* 42: 1–30.

Wrangham, R. W., and Peterson, D. (1996) *Demonic Males: Apes and the Evolution of Human Aggression.* Boston: Houghton Mifflin.

## Five: Kindness

Anderson, J. R., Myowa-Yamakoshi, M., and Matsuzawa, T. (2004) "Contagious Yawning in Chimpanzees." *Proceedings of the Royal Society of London, B (Suppl.)* 271: 468–470.

Bischof-Köhler, D. (1988) "Über den Zusammenhang von Empathie und der Fähigkeit sich im Spiegel zu erkennen." *Schweizerische Zeitschrift für Psychologie* 47: 147–159.

Boesch, C. (2003) "Complex Cooperation Among Taï Chimpanzees." In *Animal Social Complexity,* de Waal, F. B. M., and Tyack, P. L. (Eds.), pp. 93–110. Cambridge, MA: Harvard University Press.

Bonnie, K. E., and de Waal, F. B. M. (2004) "Primate Social Reciprocity and the Origin of Gratitude." In *The Psychology of Gratitude,* Emmons, R. A., and McCullough, M. E. (Eds.), pp. 213–229. Oxford: Oxford University Press.

Brosnan, S. F., and de Waal, F. B. M. (2003) "Monkeys Reject Unequal Pay." *Nature* 425: 297–299.

Caldwell, M. C., and Caldwell, D. K. (1966) "Epimeletic (Care-Giving) Behavior in Cetacea." In *Whales, Dolphins, and Porpoises,* Norris, K. S. (Ed.), pp. 755–789. Berkeley, CA: University of California Press.

Church, R. M. (1959) "Emotional Reactions of Rats to the Pain of Others." *Journal of Comparative & Physiological Psychology* 52: 132–134.

de Waal, F. B. M. (1989) "Food Sharing and Reciprocal Obligations Among Chimpanzees." *Journal of Human Evolution* 18: 433–459.

———. (1996) *Good Natured: The Origins of Right and Wrong in Humans and Other Animals.* Cambridge, MA: Harvard University Press.

———. (1997) "The Chimpanzee's Service Economy: Food for Grooming." *Evolution & Human Behavior* 18: 375–386.

———. (2000) "Attitudinal Reciprocity in Food Sharing Among Brown Capuchins." *Animal Behaviour* 60: 253–261.

de Waal, F. B. M., and Aureli, F. (1996) "Consolation, Reconciliation, and a Possible Cognitive Difference Between Macaque and Chimpanzee." In *Reaching into Thought: The Minds of the Great Apes,* Russon, A. E., Bard, K. A., and Parker, S. T. (Eds.) pp. 80–110. Cambridge: Cambridge University Press.

de Waal, F. B. M., and Berger, M. L. (2000) "Payment for Labour in Monkeys." *Nature* 404: 563.

de Waal, F. B. M., and Luttrell, L. M. (1988) "Mechanisms of Social Reciprocity in Three Primate Species: Symmetrical Relationship Characteristics or Cognition?" *Ethology & Sociobiology* 9: 101–118.

Dewsbury, D. A. (2003) "Conflicting Approaches: Operant Psychology Arrives at a Primate Laboratory." *The Behavior Analyst* 26: 253–265.

di Pellegrino, G., Fadiga, L., Fogassi, L., Gallese, V., and Rizzolatti, G. (1992) "Understanding Motor Events: A Neurophysiological Study." *Experimental Brain Research* 91: 176–180.

Dimberg, U., Thunberg, M., and Elmehed, K. (2000) "Unconscious Facial Reactions to Emotional Facial Expressions." *Psychological Science* 11: 86–89.

Fehr, E., and Schmidt, K. M. (1999) "A Theory of Fairness, Competition, and Cooperation." *Quarterly Journal of Economics* 114: 817–868.

Frank, R. H. (1988) *Passions Within Reason: The Strategic Role of the Emotions.* New York: Norton.

Gallup, G. G. (1982) "Self-Awareness and the Emergence of Mind in Primates." *American Journal of Primatology* 2: 237–248.

Gould, S. J. (1980) "So Cleverly Kind an Animal." In *Ever Since Darwin,* pp. 260–267. Harmondsworth, UK: Penguin.

Grammer, K. (1990) "Strangers Meet: Laughter and Nonverbal Signs of Interest in Opposite-Sex Encounters." *Journal of Nonverbal Behavior* 14: 209–236.

Greene, J., and Haidt, J. (2002) "How (and Where) Does Moral Judgment Work?" *Trends in Cognitive Sciences* 16: 517–523.

Haidt, J. (2001) "The Emotional Dog and its Rational Tail: A Social Intuitionist Approach to Moral Judgment." *Psychological Review* 108: 814–834.

Hare, B., Call, J., and Tomasello, M. (2001) "Do Chimpanzees Know What Conspecifics Know?" *Animal Behaviour* 61: 139–151.

Hatfield, E., Cacioppo, J. T., and Rapson, R. L. (1993) "Emotional Contagion." *Current Directions in Psychological Science* 2: 96–99.

Hume, D. (1985 [1739]) *A Treatise of Human Nature.* Harmondsworth, UK: Penguin.

Jacoby, S. (1983) *Wild Justice: The Evolution of Revenge.* New York: Harper and Row.

Kuroshima, H., Fujita, K., Adachi, I., Iwata, K., and Fuyuki, A. (2003) "A Capuchin Monkey (*Cebus Apella*) Recognizes When People Do and Do Not Know the Location of Food." *Animal Cognition* 6: 283–291.

Ladygina-Kohts, N. N. (1935 [2001]) *Infant Chimpanzee and Human Child: A Classic 1935 Comparative Study of Ape Emotions and Intelligence.* de Waal, F. B. M. (Ed.), New York: Oxford University Press.

Lethmate, J., and Dücker, G. (1973) "Untersuchungen zum Selbsterkennen im Spiegel bei Orang-Utans und einigen anderen Affenarten." *Zeitschrift für Tierpsychologie* 33: 248–269.

Leuba, J. H. (1928) "Morality Among the Animals." *Harper's Monthly* 937: 97–103.

Macintyre, A. (1999) *Dependent Rational Animals: Why Human Beings Need the Virtues.* Chicago: Open Court.

Masserman, J., Wechkin, M. S., and Terris, W. (1964) "Altruistic Behavior in Rhesus Monkeys." *American Journal of Psychiatry* 121: 584–585.

Mencius (372–289 B.C.) *The Works of Mencius.* English translation by Gu Lu. Shanghai: Shangwu.

Mendres, K. A., and de Waal, F. B. M. (2000) "Capuchins Do Cooperate: The Advantage of an Intuitive Task." *Animal Behaviour* 60: 523–529.

Nakayama, K. (2004) "Observing Conspecifics Scratching Induces a Contagion of Scratching in Japanese Monkeys (*Macaca Fuscata*)." *Journal of Comparative Psychology* 118: 20–24.

Oakley, K. (1957) *Man the Tool-Maker.* Chicago: University of Chicago Press.

O'Connell, S. M. (1995) "Empathy in Chimpanzees: Evidence for Theory of Mind?" *Primates* 36: 397–410.

Payne, K. (1998) *Silent Thunder: In the Presence of Elephants.* New York: Penguin.

Povinelli, D. J., and Eddy, T. J. (1996) "What Young Chimpanzees Know About Seeing." *Monographs of the Society for Research in Child Development* 61: 1–151.

Premack, D., and Woodruff, G. (1978) "Does the Chimpanzee Have a Theory of Mind?" *Behavioral and Brain Sciences* 4: 515–526.

Preston, S. D., and de Waal, F. B. M. (2002) "Empathy: Its Ultimate and Proximate Bases." *Behavioral and Brain Sciences* 25: 1–72.

Reiss, D., and Marino, L. (2001) "Mirror Self-Recognition in the Bottlenose Dolphin: A

Case of Cognitive Convergence." *Proceedings of the National Academy of Sciences* 98: 5937–5942.

Rose, L. (1997) "Vertebrate Predation and Food-Sharing in Cebus and Pan." *International Journal of Primatology* 18: 727–765.

Schuster, G. (September 5, 1996) "Kolosse mit sanfter Seele." *Stern* 37: 27.

Simms, M. (1997) *Darwin's Orchestra.* New York: Henry Holt and Company.

Smuts, B. B. (1985) *Sex and Friendship in Baboons.* New York: Aldine de Gruyter.

Stanford, C. B. (2001) "The Ape's Gift: Meat-Eating, Meat-Sharing, and Human Evolution." In *Tree of Origin,* de Waal, F. B. M. (Ed.), pp. 95–117. Cambridge, MA: Harvard University Press.

Surowiecki, J. (October 2003) "The Coup de Grasso." *The New Yorker.*

Tomasello, M. (1999) *The Cultural Origins of Human Cognition.* Cambridge, MA: Harvard University Press.

Trivers, R. L. (1971) "The Evolution of Reciprocal Altruism." *Quarterly Review of Biology* 46: 35–57.

Turiel, E. (1983) *The Development of Social Knowledge: Morality and Convention.* Cambridge: Cambridge University Press.

Twain, M. (1976 [1894]) *The Tragedy of Pudd'nhead Wilson.* Cutchogue, NY: Buccaneer Books.

van Baaren, R. (2003) *Mimicry in Social Perspective.* Ridderkerk, Netherlands: Ridderprint.

Watts, D. P., Colmenares, F., and Arnold, K. (2000) "Redirection, Consolation, and Male Policing: How Targets of Aggression Interact with Bystanders." In *Natural Conflict Resolution,* Aureli, F., and de Waal, F. B. M. (Eds.), pp. 281–301. Berkeley, CA: University of California Press.

Weir, A. A. S., Chappell, J., and Kacelnik, A. (2002) "Shaping of Hooks in New Caledonian Crows." *Science* 297: 981.

Westermarck, E. (1912 [1908]) *The Origin and Development of the Moral Ideas,* Volume 1, 2nd edition. London: Macmillan.

Zahn-Waxler, C., Hollenbeck, B., and Radke-Yarrow, M. (1984) "The Origins of Empathy and Altruism." In *Advances in Animal Welfare Science,* Fox, M. W., and Mickley, L. D. (Eds.), pp. 21–39. Washington, DC: Humane Society of the United States.

Zahn-Waxler, C., Radke-Yarrow, M., Wagner, E., and Chapman, M. (1992) "Development of Concern for Others." *Developmental Psychology* 28: 126–136.

### Six: The Bipolar Ape

Bilger, B. (April 2004) "The Height Gap: Why Europeans Are Getting Taller and Taller—and Americans Aren't." *The New Yorker.*

Blount, B. G. (1990) "Issues in Bonobo (*Pan Paniscus*) Sexual Behavior." *American Anthropologist* 92: 702–714.

Boesch, C. (1992) "New Elements of a Theory of Mind in Wild Chimpanzees." *Behavioral and Brain Sciences* 15: 149–150.

Bonnie, K. E., and de Waal, F. B. M. (In press) "Affiliation Promotes the Transmission of a Social Custom: Handclasp Grooming Among Captive Chimpanzees." *Primates.*

Churchill, W. S. (1991 [1932]) *Thoughts and Adventures.* New York: Norton.

Cohen, A. (September 21, 2003) "What the Monkeys Can Teach Humans About Making America Fairer." *The New York Times.*

Cole, J. (1998) *About Face.* Cambridge, MA: Bradford.

de Waal, F. B. M. (1989) *Peacemaking Among Primates.* Cambridge, MA: Harvard University Press.

———. (2002) "Evolutionary Psychology: The Wheat and the Chaff." *Current Directions in Psychological Science* 11 (6): 187–191.

de Waal, F. B. M., and Seres, M. (1997) "Propagation of Handclasp Grooming Among Captive Chimpanzees." *American Journal of Primatology* 43: 339–346.

de Waal, F. B. M., Uno, H., Luttrell, L. M., Meisner, L. F., and Jeannotte, L. A. (1996) "Behavioral Retardation in a Macaque with Autosomal Trisomy and Aging Mother." *American Journal on Mental Retardation* 100: 378–390.

Durham, W. H. (1991) *Coevolution: Genes, Culture, and Human Diversity.* Stanford, CA: Stanford University Press.

Flack, J. C., Jeannotte, L. A., and de Waal, F. B. M. (2004) "Play Signaling and the Perception of Social Rules by Juvenile Chimpanzees." *Journal of Comparative Psychology* 118: 149–159.

Frank, R. H., Gilovich, T., and Regan, D. T. (1993) "Does Studying Economics Inhibit Cooperation?" *Journal of Economic Perspectives* 7: 159–171.

Fukuyama, F. (1999) *The Great Disruption: Human Nature and the Reconstitution of Social Order.* New York: Free Press.

Galvani, A. P., and Slatkin, M. (2003) "Evaluating Plague and Smallpox as Historical Selective Pressures for the CCR5-ΔD32 HIV-Resistance Allele." *Proceedings of the National Academy of Sciences* 100 (25): 15276–15279.

Gould, S. J. (1977) *Ontogeny and Phylogeny.* Cambridge, MA: Belknap Press.

Huizinga, J. (1972 [1950]) *Homo Ludens: A Study of the Play-Element in Culture.* Boston: Beacon Press.

Kevles, D. J. (1995) *In the Name of Eugenics.* Cambridge, MA: Harvard University Press.

Lopez, B. H. (1978) *Of Wolves and Men.* New York: Scribner.

Lorenz, K. Z. (1954) *Man Meets Dog.* London: Methuen.

Malenky, R. K., and Wrangham, R. W. (1994) "A Quantitative Comparison of Terrestrial Herbaceous Food Consumption by *Pan Paniscus* in the Lomako Forest, Zaire, and *Pan Troglodytes* in the Kibale Forest, Uganda." *American Journal of Primatology* 32: 1–12.

Mech, L. D. (1988) *The Arctic Wolf: Living with the Pack.* Stillwater, MN: Voyageur Press.

Nakamura, M. (2002) "Grooming Hand-Clasp in the Mahale M Group Chimpanzees: Implications for Culture in Social Behaviours." In *Behavioural Diversity in Chimpanzees and Bonobos,* Boesch, C., Hohmann, G., and Marchant, L. F. (Eds.), pp. 71–83. Cambridge: Cambridge University Press.

Prince-Hughes, D. (2004) *Songs of the Gorilla Nation: My Journey Through Autism.* New York: Harmony.

Schleidt, W. M., and Shalter, M. D. (2003) "Co-Evolution of Humans and Canids." *Evolution & Cognition* 9: 57–72.

Scott, S., and Duncan, C. (2004) *Return of the Black Death: The World's Greatest Serial Killer.* New York: Wiley.

Shea, B. T. (1983) "Peadomorphosis and Neotony in the Pygmy Chimpanzee." *Science* 222: 521–522.

Sidanius, J., and Pratto, F. (1999) *Social Dominance: An Intergroup Theory of Social Hierarchy and Oppression.* New York: Cambridge University Press.

Silk, J. B., Alberts, S. C., and Altmann, J. (2003) "Social Bonds of Female Baboons Enhance Infant Survival." *Science* 302: 1231–1234.

Singer, P. (1999) *A Darwinian Left: Politics, Evolution, and Cooperation.* New Haven, CT: Yale University Press.

Song, S. (April 19, 2004) "Too Posh to Push." *Time*: 59–60.

Tooby, J., and Cosmides, L. (1992) "The Psychological Foundations of Culture." In *The Adapted Mind: Evolutionary Psychology and the Generation of Culture,* Barkow, J., Cosmides, L., and Tooby, J. (Eds.), pp. 19–136. New York: Oxford University Press.

Vervaecke, H. (2002) *De Bonobo's: Schalkse Apen met Menselijke Trekjes.* Leuven, Belgium: Davidson.

White, F. J., and Wrangham, R. W. (1988) "Feeding Competition and Patch Size in the Chimpanzee Species *Pan Paniscus* and *P. Troglodytes.*" *Behaviour* 105: 148–164.

Wilkinson, R. (2001) *Mind the Gap.* New Haven, CT: Yale University Press.

Wrangham, R. W. (1986) "Ecology and Social Relationships in Two Species of Chimpanzee." In *Ecology and Social Evolution: Birds and Mammals,* Rubenstein, D. I., and Wrangham, R. W. (Eds.), pp. 353–378. Princeton, NJ: Princeton University Press.

# Index

# About the Author

Frans de Waal started his career in his native Holland, where he studied biology before moving to the United States in 1981. His first book, *Chimpanzee Politics* (1982), compared the schmoozing and scheming of chimpanzees involved in power struggles with that of human politicians. Ever since, de Waal has drawn parallels between primate and human behavior, from aggression to morality and culture. His six popular books—translated into over a dozen languages—have made him one of the world's best known primatologists. He received the Los Angeles Times Book Award for *Peacemaking Among Primates* (1989) and wrote the first and only overview of bonobo behavior, *Bonobo: The Forgotten Ape* (1997).

With his discovery of reconciliation among primates, de Waal founded the field of animal conflict resolution studies. His scientific work has been published in hundreds of technical articles in journals such as *Science, Nature, Scientific American,* and publications specializing in animal behavior. De Waal has also edited nine scientific volumes.

He is C. H. Candler Professor of Primate Behavior in the psychology department of Emory University and director of the Living Links Center at the Yerkes National Primate Research Center in Atlanta. He has been elected to the National Academy of Sciences and the Royal Dutch Academy of Sciences.

With his wife, Catherine, and their four cats, de Waal lives in Stone Mountain, Georgia.

Visit the author's Web site at www.ourinnerape.com.